Semiconductor Lasers

Semiconductor Lasers

Prof. Sanjay Kumar Choudhary

AP | ARCLER
PRESS

www.arclerpress.com

Semiconductor Lasers

Prof. Sanjay Kumar Choudhary

Arcler Press
4164 Lakeshore Road
Burlington, ON L7L 1A4
Canada
www.arclerpress.com
Email: orders@arclereducation.com

ISBN: 978-1-77956-970-7 (Hardcover)

ABOUT THE AUTHOR

Prof. Sanjay Kumar Choudhary is a senior academician and experienced university administrator with a long and distinguished career in higher education. Born on 4 February 1964, he currently serves as the Vice Chancellor of Lalit Narayan Mithila University (LNMU), a position he has held since 24 January 2024. Over the years, he has played an important role in strengthening academic systems, governance, and research culture in Indian universities.

Prof. Choudhary has a strong academic background in Physics and Material Science. He holds an M.Sc. in Physics and Material Science, an M.Tech. in Surface Science and Engineering, and two Ph.D. degrees—one in Physics and Material Science and another in Electronics. He has also qualified the GATE examination in Material Science, reflecting his solid technical foundation. His main area of specialization is Material Science and Engineering.

Before joining LNMU, Prof. Choudhary served as Vice Chancellor at Munger University and T.M. Bhagalpur University, where he contributed significantly to academic planning and institutional development. He has also worked as Principal and Professor at institutions such as J.P. University, Sanjay Institute of Engineering and Management, and Bengal College of Engineering and Technology. Earlier in his career, he served as a Lecturer and Research Fellow at NIT Jamshedpur and B.I.T. Mesra, including work on an ISRO-sponsored project.

Prof. Choudhary has published 26 research papers, guided two Ph.D. scholars, and presented his work at international conferences, including events held in Germany. His academic contributions have been recognized with honors such as the Shiksha Ratna Puraskar (2010) and the Rastriya Gaurav Award (2011).

TABLE OF CONTENTS

LIST OF FIGURES

PREFACE

Semiconductor and deformed microdisk lasers have garnered interest in order to realize microlasers, and the research of semiconductor lasers has emerged as a separate field within optoelectronics. For the past 20 years, researchers have studied microcavity lasers, concentrating on their design, processing method, analysis, photonic integration, and applications. Solid-state LEDs are perfect for displays and light sources because of their brightness, whereas semiconductor laser diodes (LDs) have been utilized in everything from compact disk players to optical communications systems. However, the rareness of materials that can generate blue or green light efficiently has limited their applications. In order to replicate the visible color spectrum, full-color displays need at least three main colors: red, green, and blue. To create solid-state white-light emitting devices that are more resilient and use less energy than traditional incandescent or fluorescent bulbs, this combination is also required. Both fluorescent and incandescent bulb lamps are currently utilized as light sources for a variety of purposes. These conventional light sources are outdated, conventional glass vacuum-type light sources with low luminous efficiency and poor durability. Despite their low durability and reliability, large glass vacuum tubes were used to construct electronic circuits in the past. All electronic circuits today, however, are constructed using small and reliable solid-state semiconductor circuits. Only light sources are still manufactured using conventional technology rather than solid-state semiconductors. Compared to other traditional semiconductor devices built of GaAlAs, InGaAsP, or AlInGaP, the speed at which nitride semiconductor devices are developing is notable. The first high-brightness blue LEDs based on InGaN were introduced to the market in 1993. Then, in 1995, high-brightness blue and green LEDs with quantum well structures based on InGaN were introduced. As of 1999, these InGaN-based blue and green LEDs are utilized in a wide range of applications, including traffic lights, LED full-color displays, and others. Using the environmentally neutral substance of nitride semiconductors, they really offer higher reliability, a longer lifespan, and lower power consumption than incandescent sources, which will reduce energy consumption and safeguard natural resources. This book is suitable for all students interested in studying semiconductor lasers and their applications.

INTRODUCTION

Semiconductor lasers are widely employed in many applications, including metrology, sensors, optical communications, and optical storage. The cold atomic clock, the optical clock, high-precision interferometers, and spectroscopy are only a few examples of the ongoing advancements in laser physics and laser technology. These cutting-edge technologies place increasing technological demands on semiconductor lasers. The laser must be a high-quality single-frequency laser that operates in a single longitudinal mode and has a narrow linewidth, low noise, and high-frequency stability. The laser concept was first proposed in 1958. It was reported in 1960 that a solid-state laser and a He-Ne gas laser had been successfully demonstrated. A mode-locked laser was demonstrated shortly after the laser was developed. The mid-1960s saw the demonstration of mode locking with a saturable absorber, which is still commonly employed today. Phase locking of the laser cavity lasing longitudinal modes produces the mode-locked pulses. Many lasing longitudinal modes must be locked in phase in order to produce very short pulses. When pumped by an external laser, the dye exhibits a broad gain spectrum width and a large fluorescence spectrum width. This prompted research into dye laser mode locking, which produced extremely short pulses. Because dye lasers can produce a wide range of wavelengths and short pulses, they have been the preferred laser for many scientific studies for several decades. Due to their ease of use, semiconductor lasers became the preferred laser source for fiber optic communication systems. Today, fiber optic communication systems are in place all over the world. Short pulse generation and mode locking of semiconductor lasers at precise repetition rates were developed as a result of the realization that mode-locked semiconductor lasers are crucial for very high-speed optical time division multiplexed systems. The market need for higher-capacity transmission was aided by computer development. This is one of the main explanations for why computer processing and other important information technologies are growing at an accelerated rate in tandem with optical fiber technology. Higher capacity is still required, which is driving research into optical time division multiplexed transmission and wavelength division multiplexed (WDM)-based transmission. Research into mode-locked fiber laser operation for a variety of applications started. By altering the pump laser and the rare-earth dopant, the operating wavelength can be changed.

CHAPTER 1

INTRODUCTION TO SEMICONDUCTOR LASERS

Contents

1.1. INTRODUCTION

All semiconductor lasers (Figure 1.1) operate according to the same fundamental principles (Yamada, 2014). We can comprehend the basics of lasing if we have a better understanding of the importance and underlying technology of optical communication. The theory of stimulated emission postulates that a photon can produce other photons with the similar wavelength and phase under specific circumstances. Based on this principle, lasers produce floods of photons with identical wavelengths and phases, which generate laser light. The spectrum released by a heated black body is known as black-body radiation. The quantity of light emitted by black bodies varies with temperature, peaking at a particular wavelength (Rousseau et al., 2005). At shorter and longer wavelengths, the emission decreases to zero. As the black body's temperature rose, the peak emission moved to shorter wavelengths. Regardless of the substance, all black bodies emit light with the same spectrum when they are at the same temperature. Since photons, phonons, and particles with integral spin such as protons may have any number of particles in a given state, the Bose-Einstein distribution function is suitable for these types of particles. Particles that adhere to the Pauli exclusion principle, which states that only one particle can occupy a given energy state, are covered by the Fermi-Dirac function.

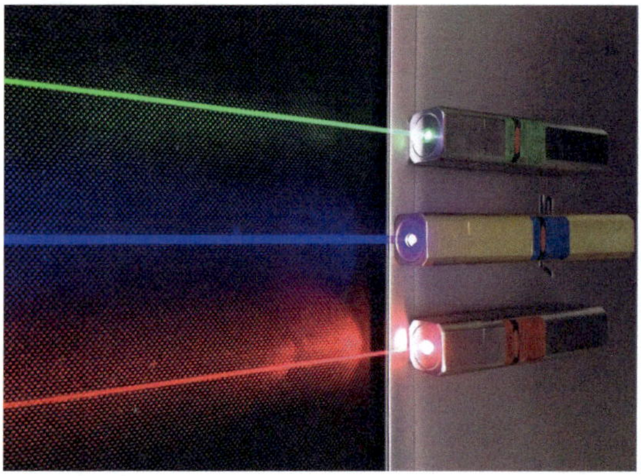

Figure 1.1. *Semiconductor lasers.*

Source: https://upload.wikimedia.org/wikipedia/commons/4/45/Lasers.jpeg, Creative Commons License.

1.2. QUANTUM

Many particles are excluded from each quantum state, not from each energy level, by this exclusion principle (Figure 1.2) (Kaplan, 1992). A collection of quantum numbers that characterize a particle is called a quantum state. In many cases, such as the sub-levels of an atomic p-orbital, there are several states with the same energy but distinct sets of quantum numbers.

We refer to these states as 'degenerate' in energy. The number of states at a given energy determines the population of electrons present at that energy. Since semiconductors have a unique crystalline structure, their bandgaps are having no specific state (Yoshikawa et al., 2007). We must calculate the density of states, or the number of photon states that can be occupied at any given energy, in order to calculate the population of photons. There must be a state in order to apply the distribution functions. For a specific physical condition or potential, these states are solutions of the Schrodinger equation. There should be more short-wavelength, high-energy photons per volume than long-wavelength, low-energy photons for a given volume. At any given time, the metal absorbs a certain number of photons as the electrons increase in energy and emits a different number of photons as the electrons relax to a lower energy level (Sturm, 1982). These rates of up and down transitions must coincide for a black body, a temperature-controlled, thermodynamic system, to be in equilibrium at a fixed temperature. The rate at which photons are absorbed and released must be equal. According to Einstein, (Figure 1.3) a black body undergoes three distinct processes: absorption, in which a photon is absorbed by the material and the material is left in an excited state; spontaneous emission, in which the material or photon relaxes to a lower energy state and a photon is emitted independently of another photon; and stimulated emission, in which the material or electron relaxes to another energy state and a photon is emitted when stimulated by another photon.

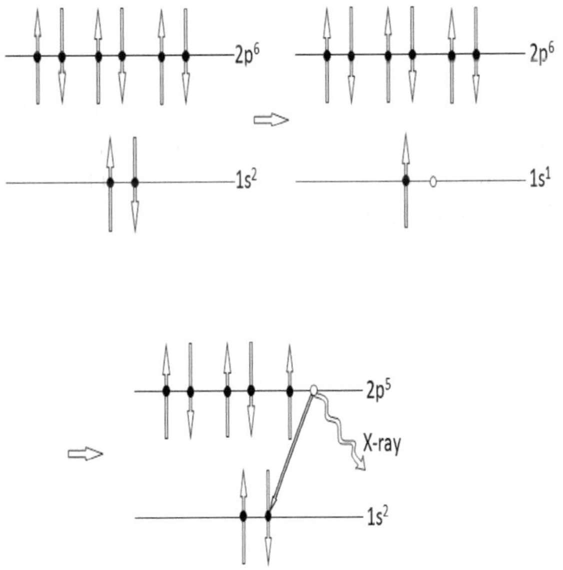

Figure 1.2. *Exclusion principle.*

Source: Kaplan, I. G. (2020). The Pauli exclusion principle and the problems of its experimental verification. Symmetry, 12(2), 320. https://doi.org/10.3390/sym12020320, Creative Commons License.

Figure 1.3. *Albert einstein.*

Source: https://upload.wikimedia.org/wikipedia/commons/thumb/1/14/Albert_Einstein_1947.jpg/960px-Albert_Einstein_1947.jpg?20220608135132, Creative Commons License.

1.3. PHOTON

A monochromatic, in-phase beam of light is lasing. A single photon triggers the emission of another photon, which in turn triggers more photons, creating an avalanche of similar photons (Zappa et al., 2007). This process is known as stimulated emission (Figure 1.4).

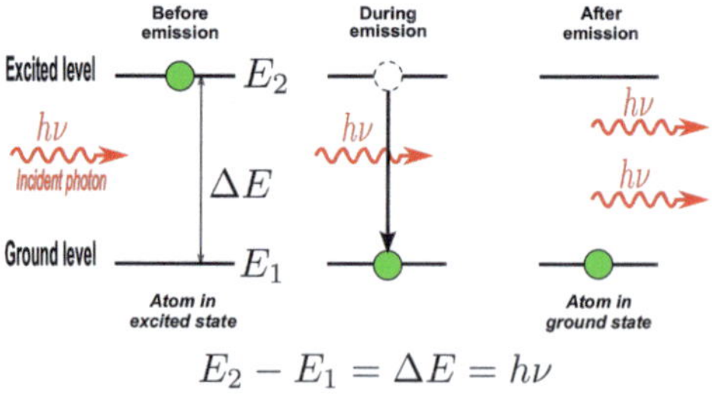

Figure 1.4. *Stimulated emission.*

Source: https://upload.wikimedia.org/wikipedia/commons/0/09/Stimulated_Emission.svg, Creative Commons License.

Since stimulated emission is the mechanism that accomplishes this, a physical scenario where the rate of stimulated emission exceeds the rate of absorption or spontaneous emission is preferred. This implies that non-equilibrium driving, usually electrical or optical, is required for lasing devices. A dominant stimulated emission cannot be obtained by thermally modulating something. An upper and lower level, where the system relaxes and emits light, and other pump level, where the system can be activated, are the levels that practical lasing systems typically consist of. The spontaneous emission rate needs to be significantly lower than the stimulated emission rate in order for lasing to occur (Yokoyama et al., 1992). Higher photon densities result in significantly more stimulated emission than lower photon densities. A larger photon density is advantageous for stimulated emission. In order to produce a high photon density inside the cavity, a laser must always have some sort of cavity mechanism based on mirrors or other wavelength-selective reflectors. Any heat-driven process is a thermal equilibrium process, and in these processes, absorption will always dominate over emission. Laser systems are driven in order to provide this non-equilibrium requirement. In semiconductor lasers, the holes and electrons are electrically injected as opposed to thermally generated (Samuel and Turnbull, 2007). The gain medium is the area of a lasing system that is in the population inversion (Figure 1.5).

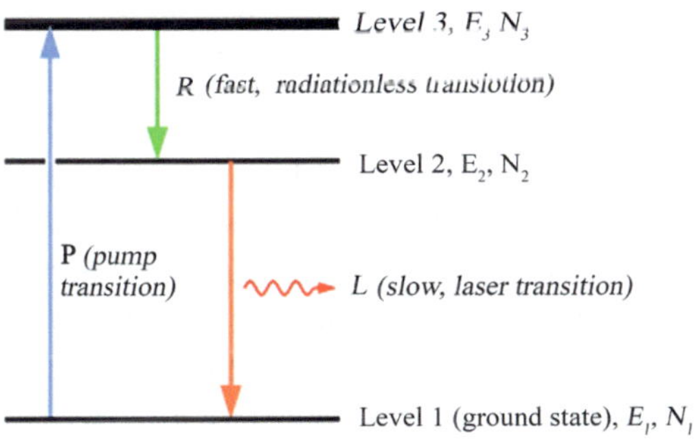

Figure 1.5. *Population inversion.*

Source: https://upload.wikimedia.org/wikipedia/commons/4/41/Population-inversion-3level.png, Creative Commons License.

A gain medium, a non-equilibrium pumping technique, and a cavity with mirrors or another mechanism to achieve a high photon density constitute every laser. To attain a high photon density, an Er or Erbium-doped fiber laser (Figure 1.6) uses a Bragg grating cavity built into the fiber as the cavity mirror, optical pumping to induce non-equilibrium, and the atomic levels of the erbium (Er) atom as the gain medium (Macêdo et al., 2019). A typical red HeNe gas laser uses high-reflectivity mirrors to define the cavity, high-voltage AC sources to electrically excite the molecules, and Ne atomic levels as the gain medium. High photon density is another prerequisite for lasing. The majority of the 1.55 μm photons are contained within the fiber laser cavity by the Bragg gratings built into the fibers (Ronnekleiv, 2001). These gratings need to have a low reflectivity at 1 μm in order to let the pump light in freely. This approach creates a device that emits a monochromatic 1.55 μm light beam when high-intensity 1 μm light is connected into the fiber.

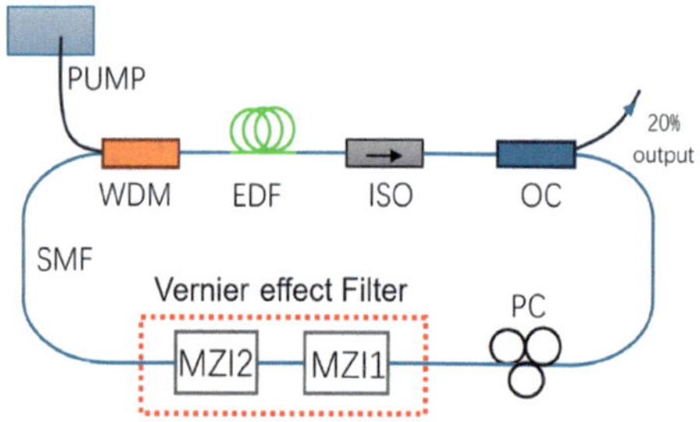

Figure 1.6. *Erbium-doped fiber laser.*

Source: Liu, Y., Xu, H., Zhu, K., Yao, Y., Suo, Y., & Zhang, L. (2024). Erbium-doped tunable fiber laser based on a vernier effect filter. Photonics, 11(10), 979. https://doi.org/10.3390/photonics11100979, Creative Commons License.

HeNe gas lasers are the conventional red lasers that are frequently utilized in optics labs (Biswas, 2023a). The HeNe molecules enclosed in the tube serve as the gain medium. A He atom is excited by the application of a high DC voltage, which produces electrons. A Ne atom then receives the energy from the He atom. A red photon is then released at $\lambda = 632$ nm as the Ne atom relaxes via radiative stimulated emission to a lower level (Svelto, 2010). The Ne atom must relax through a number of additional non-radiative stages to reach the ground state in order to be reused, even if the light has already been released. The reflectors at either end of the tube keep the photons in the cavity. The photon density inside the laser is significantly higher than the photon density directly outside the cavity since the reflectivity is usually 99% or higher (Anisimov et al., 2019). The Ne atom has many atomic levels. It is possible to cause the same system to lase in both red and green or infrared wavelengths by designing the cavity to confine photons of different wavelengths. All of these wavelengths are available in commercial HeNe lasers. After the molecule is first excited, it has sufficient time to relax from its excited state to allow for population inversion of the system. Lasing occurs when population inversion is achieved because stimulated emission dominates and the highly reflecting facets maintain a high photon density (Yang and Lyu, 2020).

1.4. BANDS

The conduction and valence bands act as the gain medium, while electrical injection is typically used as the pumping mechanism. The basic mirror mechanism in semiconductor lasers is the mirror created when a semiconductor with a refractive index of n = 3.5 cleaves and forms an interface with the air (n = 1) (Numai, 2015). Many material systems can be used to build lasers, and each type of laser has a unique set of uses. HeNe lasers are employed in optical experiments as coherent sources. In a semiconductor laser, the semiconductor serves as the gain medium. A semiconductor contains two bands: a conduction or electronic band where electrons carry current, and a valence band where holes exist and conduct current. Semiconductors are typically doped to affect their electrical characteristics. Doping is the process of adding trace amounts of atoms to a semiconductor (B – trace amounts and Si semiconductor) (Spear, 1977). The doped semiconductor has an average < four electrons since B only contains three electrons per atom in its outer shell. In the valence band, these missing electrons function as conductors. One of these charge carriers predominates in doped semiconductors; in this case, the charge conductors would be (positive) holes. The energy levels connected to one atom become the energy bands inside a crystal due to the periodicity of the crystalline array. A band gap of forbidden electron energies is formed. The average of four electrons in semiconductor compounds is sufficient to occupy the lowest energy level while leaving the higher energy levels vacant. This leads to the advantageous semiconductor characteristics of a moderate band gap and readily controllable conductivity through doping. The single band gap number does not adequately describe the complexity of semiconductor bands. Only certain semiconductors, such as GaAs and InP, which have what are known as direct band gaps, can allow light-emitting electron-hole recombinations (S. S. Li, 2006). Radiative recombination allows energy to be released as a photon when an electron and a hole recombine, eliminating both of them (Hubbard, 2016). The wavelength of light emitted by a specific semiconductor is determined by the band gap, which is the energy difference between the hole and electronic levels. When an electron in the conduction band recombines with a hole in the valence band, both are eliminated, and a photon is released.

1.5. LASER

A material's ability to function as a laser with stimulated emission improves with its ability to recombine and release light spontaneously (Mahamood, 2018). A good light emitter such as a direct band gap semiconductor can function in both spontaneous emission and stimulated emission in a laser configuration with mirrors and a non-equilibrium pumping mechanism. This is the Einstein model of stimulated/spontaneous emission, which predicts a relationship between the two coefficients of spontaneous and stimulated emission. The wavelength of light emitted from the semiconductor in telecommunications lasers is mostly determined by the band gap. We require a material with a specific band gap that emits light at a specific wavelength for semiconductor laser applications. The lattice constant of the material, or the characteristic size of the unit cell must closely match the lattice constant substrate for a positive growth. The material is grown on a semiconductor substrate. The material must have the proper bandgap for a specific wavelength of light and be approximately lattice-matched to the substrate it is grown on in order to produce a functional laser. Quantum well layers, (Figure 1.7) which are typically produced on an InP substrate, are the layers used to create communications lasers (Amann and Ortsiefer, 2006). It is crucial that the layer lattice constant value be near 5.8686. Å, regardless of wavelength. The ability to create good heterostructures of the four fundamental elements, In and Ga being freely interchangeable, and As and P being freely interchangeable, is what makes this material system useful.

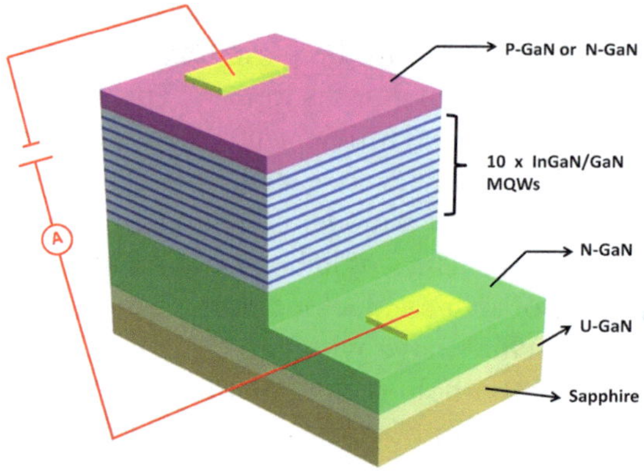

Figure 1.7. *Quantum well layers.*

Source: Yang, H., Ma, Z., Jiang, Y. et al. (2017). The enhanced photo absorption and carrier transportation of InGaN/GaN quantum wells for photodiode detector applications. Sci. Rep., 7, 43357. https://doi.org/10.1038/srep43357, Creative Commons License.

A wide range of band gaps and lattice constants can be found in the quaternary compounds of $In_xGa_{1-x}As_yP_{1-y}$ (Othman et al., 2010). There is a combination of InGaAsP that will fit both the lattice constant and the required band gap. The bandgap of InP-based lasers overlaps both 1.55 and 1.3 µm, which are the low loss and low dispersion windows for optical fiber, respectively (Sneh and Doerr, 1999). This makes them useful for telecommunications applications. GaAs-based lasers (Figure 1.8) are employed as a crucial amplifier part to produce lasers with a wavelength of about 1 μm (Pessa et al., 2003). By creating heterostructures or combinations of binary compounds, a variety of materials with various bandgaps or wavelengths can be produced. The total number of group III and group V atoms in the compound is always limited to an even number. Since thin films are deposited on a substrate to create quantum wells, they are crucial to the production of the lasers.

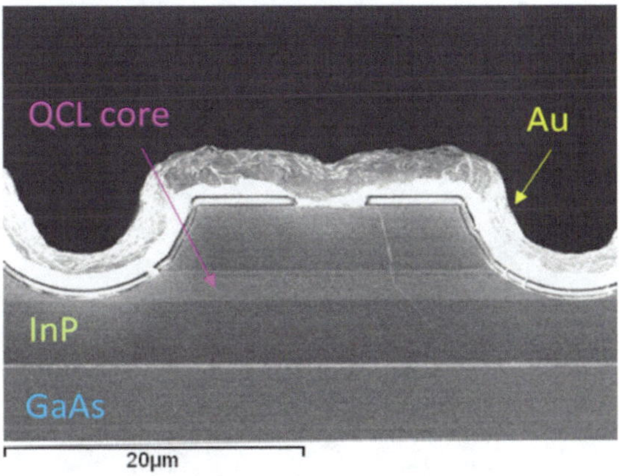

Figure 1.8. *GaAs-based lasers.*

Source: Slivken, S., & Razeghi, M. (2022). High power mid-infrared quantum cascade lasers grown on GaAs. Photonics, 9(4), 231. https://doi.org/10.3390/photonics9040231, Creative Commons License.

It is crucial to consider the effects of thin film deposition on a substrate on both its electrical and physical characteristics. The basic size of a semiconductor unit is known as the lattice constant (Giesecke, 1966). The material experiences strain when the thin film lattice constant differs from that of the substrate, which it is growing on. A layer of material placed on top of a material with a different lattice constant is strained, much like a spring is when it is compressed or extended, exerting force to return to its desired dimension. Growing a stretched layer too thick will cause the atomic bonds to break, resulting in dislocations. The degree of lattice mismatch in the material determines the critical thickness, or the maximum thickness a strained layer can grow without experiencing dislocations. Because a low defect density is essential for optimum laser performance, strain and critical thickness are crucial for producing these thin layers utilized in lasers. One type of material defect is strain-induced dislocations. Recombination centers are created by defects such as excess or missing atoms, which lead to carriers recombining and producing heat instead of light. These semiconductor thin films must closely match the substrate lattice constant in order for devices to function as emitters. On a semiconductor wafer known as a substrate (InP), the active semiconductor layers are formed (Hjort, 2004).

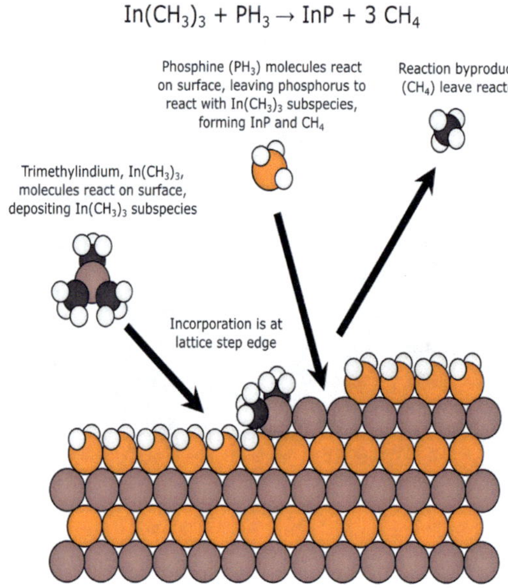

Figure 1.9. *Metallorganic chemical vapor deposition.*

Source: https://upload.wikimedia.org/wikipedia/commons/f/f4/MOCVD_process.svg, Creative Commons License.

Atoms are deposited onto the preexisting substrate using semiconductor growth techniques like metallorganic chemical vapor deposition, or MOCVD (Figure 1.9); molecular beam oxide, or MBE, and bond to the preexisting layers. Since they function as non-radiative recombination sites that compete with radiative recombination for carriers, these dislocations are problematic for lasers. Layers often contain strains of 1% or higher. It may be advantageous to apply a small amount of strain to enhance the device speed or other characteristics. It requires more energy to keep atomic layers squeezed into their non-equilibrium structure the more of them are stacked together. Before dislocations begin to manifest, these thin layers can only be developed to a specific thickness. This thickness is crucial for lasers and is known as the critical thickness. Quantum wells, which are thin (100 Å) layers of one material layered between other thin layers of material, are the building blocks of quantum well lasers (Fox and Ispasoiu, 2017). Because these layers are typically not exactly lattice-matched to their substrate, it is critical to understand the strain and material limitations on the maximum thickness at which these layers can stack. It is energetically beneficial to have a limited broken bonds in one layer if there are many atoms in a thick strained layer. This allows for the growth of a relaxed layer with its equilibrium lattice constant not matching the substrate. If there are only a few atoms in a thin layer, they will be strained and match up to the substrate. The thermodynamic equilibrium of minimal energy serves as the foundation for this critical thickness model, which compares strain and dislocation energies (Fang et al., 2022). After 50 or 100 layers of atoms have been produced, there is a kinetic barrier to moving to a dislocation, starting with a few stressed layers. Layers thicker than the critical thickness can typically be developed without dislocation. Compressive strain would be 0.35% and the critical thickness would be 800 Å for an InGaAs layer with a lattice constant of 5.67 Å formed on a GaAs substrate with a lattice constant of 5.65 Å (Klotzkin, 2014b). Since this strain is cumulative, 800 Å of InGaAs can be generated by growing alternating layers of GaAs and InGaAs on a GaAs (Fleischmann et al., 2001). Quantum wells also employ a technique that enables the growth of as many distinct thin layers as needed. Compressive and tensile stressed layers are paired in strain compensation, which is utilized in multiple quantum well lasers. The tension is eliminated, allowing for the growth of extremely thick layers. The composition of the material determines properties like the band gap and lattice constant, and thin films have additional limitations dependent on the amount of strain the material can withstand, even though they may concentrate electrons and holes to very high densities and facilities lasing. The band gap, which is the smallest distance between the highest valence energy level and the lowest electronic energy level, is typically what researchers are most interested in (Smith and Nie, 2010). The majority of carriers will be in their lowest energy state, where recombinations will occur, because electrons (and holes) will remain at this state. The majority of the holes will also be near the zone center, where the valence band minimum (hole) energy is precisely above the conduction band minimum (Rediker, 1965). For a laser material, this is very significant. Qualitatively, momentum is a property shared by electrons and holes, and like energy, momentum must be preserved in some way during an interaction. The de Broglie relation p = hk provides the momentum connected to an electron, hole, or photon in a crystal (Holland, 1993). A net change in momentum, hΔk, and energy equivalent to the band gap results from an electron switching from a state in the conduction band to a state in the valence band during a recombination event. The photon that is released absorbs the energy, but it has very low momentum. In a radiative recombination, momentum must either be conserved in some other way or Δk must be zero (Vardi et al., 1997). This radiative recombination is significantly less likely when three elements are involved i.e., an electron, hole, or phonon. The semiconductor must be a direct-band-gap material, meaning that the minimum energy in the conduction band must be immediately above the minimum (hole) energy in the valence band, in order for Δk to equal zero.

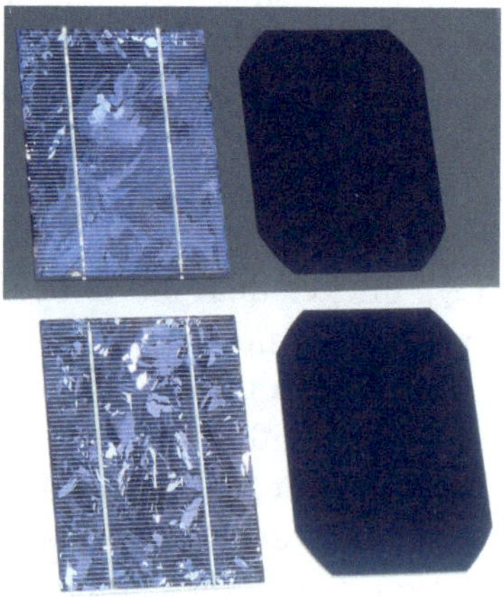

Figure 1.10. *Semiconductor crystal.*

Source: http://upload.wikimedia.org/wikipedia/commons/7/71/Comparison_solar_cell_poly-Si_vs_mono-Si.png, Creative Commons License.

This implies that k = 0 for both holes and electrons. Direct band gap semiconductors, which have valence band and conduction band energies with minima at the same k-value, include GaAs, InP, and the majority of their heterostructures, including InGaAsP (Mishra and Singh, 2008a). The semiconductor Si is not a direct band gap material. It is evident that at k = 0, the valence band (electron) minimum and the conduction band minimum do not overlap (Morkoç et al., 1994). Therefore, regardless of how advanced or affordable Si wafers become, Si will never be a good laser or light-emitting device. Si can be a good light detector. In addition to the creation of electron-hole pairs, lattice vibrations in the atoms are also produced or absorbed as a result of the interaction of phonons, or lattice vibrations, which conserve momentum when light is absorbed. Because absorption is a more efficient process than recombination, Si can detect light without having the ability to easily produce it. Phonons are helpful in permitting certain absorptions and recombinations between carriers with varying k-values. A semiconductor crystal (Figure 1.10) is made up of several atoms that are linked together, but at temperatures higher than zero, each atom vibrates slightly from its equilibrium position. Atomic vibrations rise with increasing temperature. In many carrier-light interactions, excess momentum is absorbed by these lattice vibrations. Phonon interaction facilitates an absorption event in Si. Si absorbs a photon at 700 nm, transitioning with a Δk of roughly $2/3\ \pi/a$ (Plekhanov, 2012). Either optical phonon emission, which results in an absorption energy 30 meV below, or optical phonon absorption, which results in an absorption energy 30 meV above 1.77 eV, absorb the change in system momentum (Gamaly and Rode, 2013).

1.6. QUANTUM WELLS

A small piece of material with a lower band gap positioned between two other materials with a greater band gap is called a quantum well (Adams et al., 1999). The carriers i.e., holes and electrons are confined by these energy walls to remain primarily within the well. Carriers are restricted in confining potentials created by these semiconductor quantum wells. The density of electrons and holes in the same area is significantly higher because they are constrained by the surrounding energy barriers. The realization of high-performance semiconductor lasers depends on this increase in carrier density (Liu et al., 2015a). The significance of quantum wells in contemporary semiconductor lasers cannot be exaggerated. Creating a functional laser at high temperatures using bulk semiconductor material is quite challenging. Higher current densities would be needed to laser light and unconfined carriers. All of the performance characteristics are significantly better and the carrier density in the quantum well is significantly higher than in a bulk p–n junction with the same current input. When a substance has optical gain, more light exits the material than enters when light is fired on it or through it. Although light absorption is frequent, optical gain has a significant role in technology. Based on optical gain, the erbium-doped fiber amplifier can magnify signals by a factor of 1000, enabling optical transmission over thousands of miles (Peterka and Vojtěch, 2024). The following formula describes optical gain and absorption.

where P_0 is the initial optical power, P is the final power, and the gain is denoted by g, is in length^{-1} units and is ($^{-}$ve) for absorption and ($^{+}$ve) for actual gain. Gain and absorption are commonly given in units of cm^{-1} in laser applications.

No two electrons may have the same quantum numbers or occupy the same quantum mechanical state, according to the Pauli exclusion principle. The number of states accessible to electrons or holes at a specific energy is provided by the density of states formula of a semiconductor. A bulk semiconductor density of states grows quadratically with energy (Ohtsubo, 2017). A quantum well has a constant two-dimensional density of states. The third-dimension sub-bands produce a density of states versus energy that resembles a staircase. Because it produces many carriers at the same energy, a quantum well sudden increase in density of states is ideal for lasers.

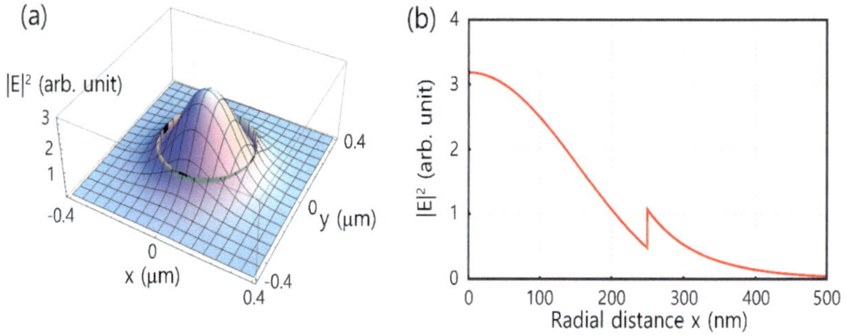

Figure 1.11. *Optical wave (a & b).*

Source: Ihn, Y. S., Kim, Z., & Lee, S. Y. (2020). Optical wave guiding and spectral properties of micro/nanofibers used for quantum sensing and quantum light generation. Applied Sciences, 10(2), 715. https://doi.org/10.3390/app10020715, Creative Commons License.

Semiconductor lasers are nearly always quantum wells or smaller in size, and threshold current densities are significantly lower (Bimberg et al., 2000). The product of the Fermi function and the density of states determines the number of carriers in a band at a specific energy. The semiconductor is not in thermal equilibrium when electrical or optical injection is present. Distinct quasi-Fermi levels can be used to characterize the populations of electrons and holes (Pirouz, 2013). For stimulated emission to predominate over absorption, the lasing energy E_λ must be smaller than the split between the quasi-Fermi levels. The density of states, the occupancy of holes and electrons, a proportionality constant, and a broadening factor affect optical gain. A Gaussian or Lorentzian expression with a linewidth is used to describe the broadening factor. R (power reflectivity) is roughly 0.3 for common semiconductor lasers (Wenzel et al., 2021). 30% of the incident is reflected back into the cavity by the integrated reflectors in these cleaved laser bars. In these structures, lasing can be achieved with this reflectivity. Commercial devices typically have their facets coated with dielectric layers after production in order to improve or decrease their reflectivity at particular wavelengths. When the photon density is high, the feedback between the gain and the density is significant. An electron and a hole must be removed in order to produce a photon. The gain drops as the photon density rises and the hole and electron densities fall. The laser is an optical amplifier with feedback. The phrase steady state refers to a situation in which nothing changes over time; the injected current, carrier density, and photon density inside the cavity appear the same as they did or will in 15 minutes (Shayeganrad, 2020). When a laser is said to have "unity round trip gain," it means that the power of the optical wave should be equal to its initial level after bouncing back and forth between the cavities; this includes power that leaks out of the facets (Klotzkin, 2014a).

1.7. OPTICAL WAVE

The optical wave (Figure 1.11) would be larger after one round. The wave would get bigger as it circled around again. The gain would eventually decrease when the photon density in the cavity increased more, depleting the electrons and holes (Yamamoto et al., 1992). The carrier density would then increase to reach unity round trip gain if the gain was less than equilibrium. Due to the association between photon density and carrier density, the gain value must be such that the laser remains in a steady state. The cavity characteristics, including facet reflectivity, determine the precise value of the equilibrium gain 'g' (Le et al., 1997). To attain unity round trip gain, the gain is fixed in a lasing cavity and is constant. Similar to electronics, the intrinsic material gain establishes boundaries, but the closed-loop gain is crucial in determining device characteristics. The gain in the cavity cannot vary with current density. While the carrier density inside a laser cavity is fixed above threshold and unaffected by the photon density, the photon density inside the cavity will fluctuate. Free-carrier absorption is the important extra absorption component in laser design (Bulashevich et al., 2007). Excitonic absorption is a subclass of band-to-band absorption that is frequently observed at very low temperatures or at higher temperatures in highly pure semiconductors and quantum wells. An exciton is a pair of electrons and holes that, at low temperatures, create a Coulombic attachment that reduces their respective energies. This coupled electron-hole pair is an exciton, which is eliminated when it is absorbed by a photon. Excitonic absorption is the cause of additional absorption peaks observed near a semiconductor band edge. A carrier (electron or hole) is excited when a photon strikes a semiconductor. In its band, this electron or hole is promoted. Through interaction with the lattice and other carriers, the excited carrier returns to its equilibrium location within the band. The doping density affects this process; the higher the doping density, the more probable it is that this absorption process will occur (Zuo et al., 2016). The confined area surrounding the quantum well is typically left undoped.

1.8. ABSORPTION

More free carrier absorption occurs in heavily doped regions for relatively low energy photons due to the wavelength dependence. High power operation devices maintain a low absorption value, pump lasers with several 100 mW of power usually have absorption losses between 2 to 5/cm (Koechner, 1996). The numbers for high speed modulated telecommunications devices are closer to 20/cm (Ahmed et al., 2014). The confining heterostructure region is lightly doped to minimize absorption losses because this process is dependent on the density of carriers in the area close to the semiconductor. Better conductivity and reduced heat dissipation are two benefits of high doping, but there are some problems.

Improved modulation performance may result from higher p-doping in the active area. Light can be absorbed anywhere a carrier may be absorbed and to change energy states. In semiconductors, impurities that trap carriers can also act as absorption sites. Low energy absorption from impurities to conductor or valence bands is common. Since there are fewer impurities in high-quality lasing material and the absorption energy is lower than the lasing energy, these mechanisms are not very significant in lasers. The trap-based recombination is frequently insignificant since good lasers usually have very low defect densities. Bimolecular recombination is the used to describe devices with shorter wavelengths i.e., 980 nm (Cowan et al., 2013). Auger recombination is important for devices with longer wavelengths, and it is also important at higher carrier densities. In the majority of semiconductor edge-emitting laser applications, the two facet reflectivities are not equivalent.

The wafer is physically cleaved to create the mirrors in edge-emitting Fabry-Perot lasers (Zappe, 2004a). Using a diamond-tipped tool, the wafers are scribed on an edge before breaking; the break propagates along the crystal planes to create a perfect dielectric mirror between the semiconductor and air. Since these mirrors are symmetrical when they are constructed, half of the light would leave the cavity on one side and half on the other. When doing this, it is critical to align the scribe and cleave marks with the wafer plane of cleavage. For commercial purposes, it is preferable that the majority of the light leave one facet to be coupled into an optical fiber. To alter the reflectance, dielectric coatings are typically applied to the facets. A Fabry-Perot laser design (Figure 1.12) has a front facet reflectivity of 10% and a rear facet reflection of around 70% (Liao et al., 2009). A minute quantity of light leaves the rear facet of the laser, whereas the majority leaves the front facet. In order to enable active control of the output laser power, the rear facet light is frequently connected to a monitor photodiode within the package. The qualities of the laser can be effectively controlled with these coated facets. It is evident that as facet reflectivity rises, the necessary cavity gain falls. So, raising the reflectivity of the facet coating can lower the threshold current needed.

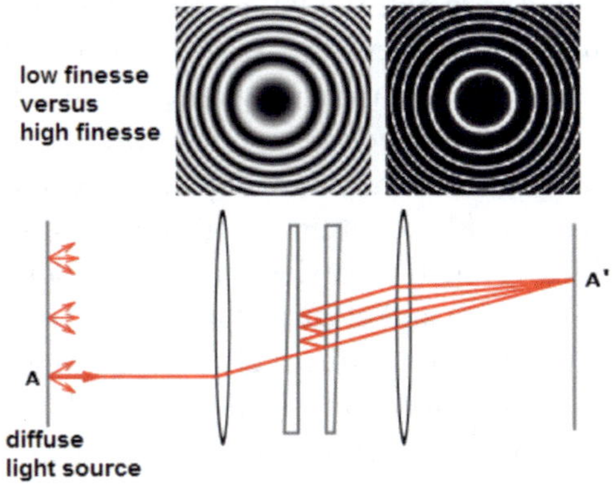

Figure 1.12. *Fabry-Perot laser.*

Source: https://upload.wikimedia.org/wikipedia/commons/5/56/Fabry_Perot_Interferometer_-_diagram.png, Creative Commons License.

1.9. ANALYSIS

The material limits the laser properties first, followed by the structure. Broad-area samples that are evaluated using pulsed current sources are employed for material analysis. A p-doped zone on one side, an undoped region of quantum wells and barriers in the central region, known as the active region, and an n-doped region on the other side make up semiconductor laser diodes (Zappe, 2004b). One side is used to inject electrons, while the other side is used to inject holes. The active area is where holes and electrons build up. The entire structure is described by a Fermi level. The number of mobile electrons or holes in the band is determined by the separation between the valence and conduction bands, respectively, and the Fermi level. The semiconductors appear to be either n-doped or p-doped when they are far from the junction of the n- and p-regions (Huebener, 2016). On the n-side, the Fermi level is close to the conduction band, the electron density is roughly equal to the dopant density, and the hole density is $n_i^2 = N_D$ (Milnes, 1980). Since dopant atoms provide the majority of their mobile electrons, they are electrically neutral. A dopant atom with a fixed positive charge is left by each mobile electron with a negative charge. It has no net charge and is electrically neutral. There are not many mobile carriers in the intermediate region, where the Fermi level is distant from the valence and conduction bands, but the dopant atom stationary charge is still present.

This is referred to as the depletion region. The space charge remains when the free electrons and holes recombined and disappeared at the intersection of the electron-rich n-doped side and the hole-rich p-doped side (Ma and Chen, 2017). There is a small area where the semiconductor changes from being quasi-neutral, with no net charge, to having an electric field and significantly fewer mobile carriers at the intersection of these two zones. The Debye length is relatively small, even for relatively low dopant concentrations (Brenier and Boulon, 2001). It is reasonable to assume that the quasi-neutral zone and the depletion region will intersect. The carriers bounce randomly off of atoms in a crystal lattice in both diffusion and electrical mobility. The electric field causes displacement between collisions, effectively rearranging the carrier path of travel; in diffusion, the random motion is always random but accumulates to move the carriers from high-concentration areas to low-concentration areas. Minority carriers (holes) from the n-side of the quasi-neutral zone that fall into the space charge region will move toward the p-side, while minority electrons from the p-side will move toward the n-side (Mishra and Singh, 2008b). In contrast to the diffusion current, the drift current flows in the opposite direction. The net current is zero at equilibrium.

A built-in voltage between one side and the other, as well as a region entirely depleted of mobile charges sandwiched between quasi-neutral regions that are charge neutral, are the basis for various formulas that can be derived for other dopant formulations. The lightly doped side of the junction has a wide depletion layer due to overall charge neutrality. The depletion layer width on the p-side will be 10 times greater than on the n-doped side if $N_D = 10N_A$ (Liu et al., 2010). We can assume that all of the depletion width occurs on the lightly doped side if one doping is larger than the other. The central section of a laser with an undoped active region has a diode that is undoped. Given that it has comparatively few mobile charges, the undoped central region appears to be a part of the depletion region.

The voltage drop over the undoped zone absorbs the majority of the built-in voltage, whereas depleted n- and p-layers are visible near the margins of the doped active regions. A positive current moves from the p-side to the n-side as a result of the barrier height decreasing due to its forward bias (Biswas, 2023b). The reduction of the barrier height also results in a reduction of the depletion layer width. Current begins to flow when this bias is applied to the p-side. It must be the diffusion current that rises with increasing voltage as it is the current that moves from the p-side to the n-side. Approximately the same amount of minority carriers becomes drift currents after becoming trapped in the depletion region, regardless of the size of the region.

1.10. HOLES

More holes are injected into the depletion region as a result of the bias. These excess carriers are present in the periphery of the p-side quasi-neutral area, assuming minute recombination. These excess minority carrier holes recombine with the majority carrier electrons in the quasi-neutral region. The intrinsic carrier density with that doping density is the carrier density far from the junction. The quasi-Fermi levels separate and an excess of minority carriers begins to appear close to the depletion region edge (Wenzel and Zeghuzi, 2017). In order to preserve quasi-neutrality, there are also the equal number of surplus majority carriers. There are more electrons and holes than there would be in equilibrium across the depletion zone. The excess carriers on each side are injected across the depletion area and show up on the opposite side since it is anticipated that the carrier density is still too low for recombination. The reciprocal of the slope at each location is this differential resistance. The splitting of the quasi-Fermi levels occurs as the voltage rises (Caprioglio et al., 2019). The quasi-Fermi levels in a laser are clamped above the threshold, and all additional carriers introduced into the active zone leave as photons over the threshold (Dutton and Yu, 1993). The differential resistance is clamped as a result of the clamping of the quasi-Fermi levels. Because of the metal contact resistance and the ohmic resistance across the p- and n-side of the active region, this differential resistance is actually no longer a diode resistance; rather, it reflects the resistance. The differential resistance curves of a laser diode and a typical diode differ drastically. A laser resistance decreases and remains constant at the threshold, reaching a value equivalent to the resistance. A diode differential resistance is continuously dropping. At threshold, the laser diode essentially has a clamped band structure rather than being a diode. Getting a good metal-to-semiconductor contact is an obstacle that was initially linked to Schottky (Evstigneev, 2022). A simple metal has many states below and above the Fermi level, unlike a semiconductor. All states below the Fermi level are occupied, while all states above the Fermi level are unoccupied. The electron affinity, or qX, of a semiconductor is a value that is comparable to but distinct from the metal work function. The energy required to extract an electron from a semiconductor is represented by the electron affinity, which is the energy gap between the conduction band and the vacuum level. For semiconductors, this is the important material constant.

1.11. SEMICONDUCTOR

Si has an electron affinity of 4.35 eV. Due to its high work function, Pt is typically utilized for p-metal contact (Lu et al., 2012). Ti and NiGe alloys, both of which have relatively low work functions, are used to make contact with the n-side. The band structure at the semiconductor surface, where the metal is deposited, differs from that in the semiconductor bulk, but this is ignored by junction theory. Another crucial aspect of semiconductor laser functioning is their electrical properties. Ohmic heating is reduced in contacts with low resistance. At their core, semiconductor lasers are p-n junctions. A depletion area is created by the p-n junctions, where stationary depletion charge remains after mobile electrons and holes recombine. An electric field and an inherent voltage between one side of the junction and the other are created by the depletion charge. The quasi-neutral area, where the net charge is zero, is located on either side of the depletion region (Bisquert et al., 1999). It is assumed that the depletion region and the quasi-neutral region have borders. A drift current flows from the n-side to the p-side due to the electric field across the depletion zone, and a diffusion current flows from the p-side to the n-side. In equilibrium, these currents are balanced. The built-in voltage is decreased by applied forward bias. While the diffusion current magnitude grows exponentially, the drift current magnitude stays constant. The amount of excess carriers injected into each side of the quasi-neutral zone depends exponentially on voltage, a Fermi level split across the junction. As these excess carriers disseminate into the quasi-neutral area, they recombine. P-N junctions and lasers are different. The diode ideality factor is usually closer to two than one since lasers have a large recombination current. The quasi-Fermi level in lasers is clamped above threshold.

CHAPTER 2

NITRIDE SEMICONDUCTOR LASERS

Contents

2.1. INTRODUCTION

Group-III nitride-based semiconductors (Figure 2.1) have recently become the primary material used to produce blue laser diodes and blue LEDs (Hardy et al., 2011). A continuous and direct bandgap semiconductor alloy that spans ultraviolet to blue/green/yellow wavelengths is from nature environment for the (In_xGa_{1-x}), $Al_{1-y}N$ alloy system.

The direct bandgap semiconductor is the effective way to produce light from electricity. The development of blue InGaN LEDs has revolutionized LED technology (Shen, 2015).

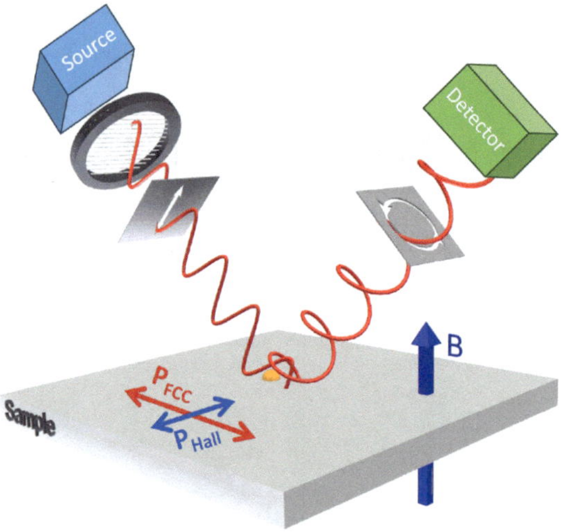

Figure 2.1. *Group-III nitride-based semiconductors.*

Source: Armakavicius, N., Kühne, P., Papamichail, A., Zhang, H., Knight, S., Persson, A., Stanishev, V., Chen, J.-T., Paskov, P., Schubert, M., & Darakchieva, V. (2024). Electronic properties of group-III nitride semiconductors and device structures probed by THz optical hall effect. Materials, 17(13), 3343. https://doi.org/10.3390/ma17133343, Creative Commons License.

A number of significant developments in the production of materials were necessary for the advancements in GaN-based LED materials. For many years, research on Group-III nitrides and poor crystal quality was delayed due to lack of p/n junctions. Rejuvenating the III-V nitride system is generally attributed to Akasaki who worked on p-doping in GaN and Nakamura application of InGaN/GaN double heterostructure (DH) in LEDs in 1994 (Figure 2.2) (Monemar, 2008).

The current development of blue lasers comes more than two decades after the first optical pumped stimulated emission was detected in GaN crystal and the first LEDs were created. Significant advancements in high-brightness LED displays were made in the late 1980s. The production of high-efficiency AlGaInP, AlGaAs, and AlGaInN alloys is a remarkable advancement. Following Akasaki's initial discoveries of p/n junctions in 1989, the efficiency of p/n GaN LEDs increased quickly (Akasaki, 2007). The higher efficiency provided by heterojunction-based LEDs was the main cause of this. Many new industries for LEDs have been made possible by advancements in GaN and GaAs-based LEDs. The development of a powerful blue light source has long been a goal for academics working on solid state devices. Now, it is possible to use three LEDs to any visible color or even create white LEDs by combining a single blue LED with phosphors.

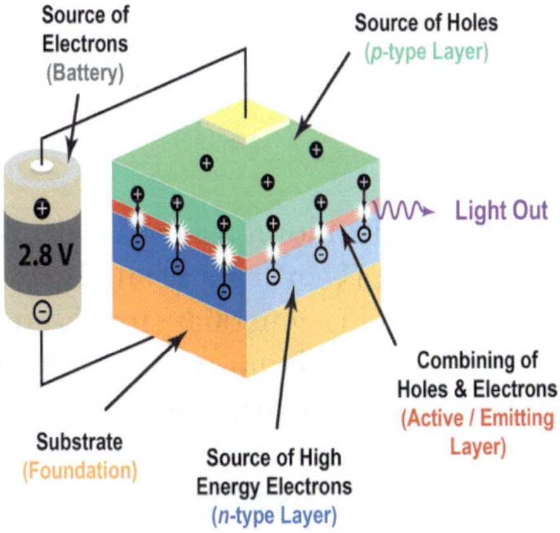

Figure 2.2. *Double heterostructure.*

Source: Nakamura, S. (2015). Background story of the invention of efficient blue InGaN light emitting diodes (Nobel Lecture). Annalen Der Physik, 527(5, 6), 335–349. https://doi.org/10.1002/andp.201500801, Creative Commons License.

The solid-state characteristic of semiconductor devices results in extremely high reliability, which makes this feasible. Full-color and white LEDs can be found in a range of devices, including scanners, outdoor TVs, traffic signals, and backlighting for cars.

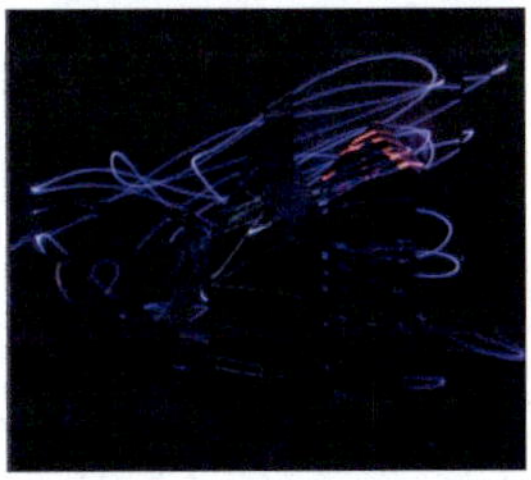

Figure 2.3. *Blue laser diodes.*

Source: https://upload.wikimedia.org/wikipedia/commons/5/56/Blacklaser2.jpg, Creative Commons License.

The fabrication of blue laser diodes (LDs) (Figure 2.3) for exceptionally high-density optical storage devices is another potentially significant use for GaN. Short wavelength laser diodes could result in a 4–8 fold increase in capacity because the storage density of optical compact discs (CDs) and digital video discs (DVDs) is inversely proportional to the square of the laser wavelength (Morkoç, 1999). To commercialize GaN for optical storage, energy-efficient lighting, communications, printing, TV projection, and surgery, more research is required.

2.2. LED

GaN LEDs and LDs should be used in more areas where dependable and affordable lighting is required as costs rise (DenBaars et al., 2013). Understanding the science underlying light emission in semiconductors is crucial to understanding GaN LEDs. Under the forward bias circumstances, a type of diodes known as light emitting diodes (LEDs) can produce spontaneous radiation. A p/n junction is created in semiconductor materials used to make LEDs. The crucial method for stimulating semiconductor materials is injection electroluminescence (EL). The p-type semiconductor receives an injection of electrons under forward bias circumstances, whereas the n-type material receives an injection of holes. Near bandgap radiation is produced when these minority carriers recombine with the majority carriers at the p/n junction.

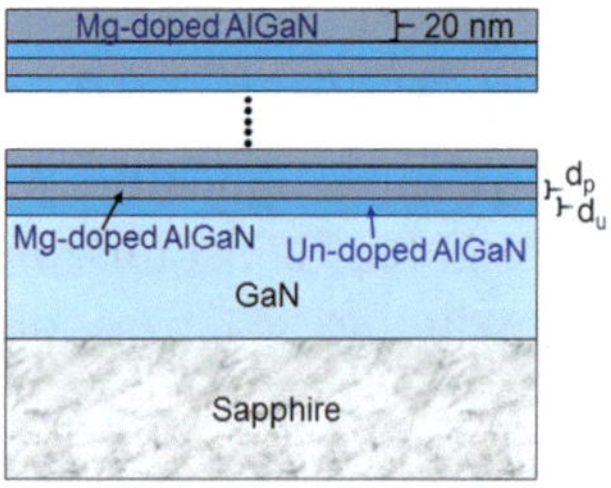

Figure 2.4. *Mg-doped GaN.*

Source: Chen, C.-C., Lin, Y.-R., Lin, Y.-W., Su, Y.-C., Chen, C.-C., Huang, T.-C., Wu, P.-H., Yang, C. C., Mou, S., & Averett, K. L. (2021). Improvement of p-Type AlGaN conductivity with an alternating Mg-doped/un-doped AlGaN layer structure. Micromachines, 12(7), 835. https://doi.org/10.3390/mi12070835, Creative Commons License.

Mg-doped GaN (Figure 2.4) is used to create the p-type material in GaN, while silicon doped GaN is used to create the n-type material. Before 1989, there were only substances that emitted radiation that covered the green through infrared spectrum (Ribeiro and Crowley, 2007). In 1962, General Electric created the first visible LEDs, which were red GaAsP emitters (Shchekin and Craford, 2016). Since blue was one of the primary colors i.e., red, green, and blue, full-color emission was not possible. The full-visible color spectrum is now available for the first time due to the development of the GaN technology. Both direct and indirect bandgap behavior are characteristics of semiconductor LED materials, which fall into distinct types of band structure. Electrons in the conduction band minima and holes in the valence band maxima of a direct bandgap semiconductor have the same momentum. For the bandgap semiconductor, the likelihood of a band-to-band radiative transition is high because recombination will conserve momentum. Momentum is not conserved in an indirect bandgap semiconductor because the electrons in the conduction band minima and the holes in the valence band maxima have different momentum. Before recombining with a hole in an indirect semiconductor, an electron would need to shift its momentum. For the electron to alter its momentum and a transition to take place in the indirect gap semiconductor, a phonon must take part in the recombination event. As a result, the indirect gap semiconductor band-to-band transition recombination probability is reduced. When comparing direct bandgap GaN to indirect SiC LEDs, this becomes clear. Compared to SiC LEDs, which have quantum efficiencies of 0.02%, GaN LEDs have quantum efficiencies of up to 12% (Edmond et al., 2004).

2.3. ELECTRON-HOLE RECOMBINATION

There are two types of electron-hole recombination: radiative and non-radiative (Cheng et al., 2022). The two recombination paths can be thought of as parallel processes taking place across the semiconductor energy gap. A LED light output can be increased by increasing the likelihood of radiative recombination and decreasing the likelihood of nonradiative recombination channels. Non-radiative recombination in semiconductors is caused by a number of processes, including multiphonon emission at deep impurity sizes, auger recombination, and recombination at defect sites. In semiconductors, excitonic recombination, interband transitions, and recombination through impurity centers are the main causes of radiative transitions (Burdov and Vasilevskiy, 2021). The amount of light produced for the injected current is the indicator of LED efficiency. The majority of carriers recombine via the radiative pathway due to GaN direct bandgap, resulting in quantum efficiencies of up to 10% (Chichibu et al., 2018). The external quantum efficiency is essentially the ratio of photons out to injected electrons. Three requirements must be met for a material to be an effective emitter of visible light. They are (a). it must have an effective radiative pathway; (b). it must have an energy gap of a suitable width; and (c). it must be doped both p- and n-type with low resistivities. III-V semiconductors and their alloys are the significant materials that has achieved technological achievement in satisfying these requirements. Lumens per watt is the unit of measurement for luminous efficiency. The energy conversion efficiency of the light source in transforming electrical power into visible light, or luminous flux, is referred to as luminous efficiency.

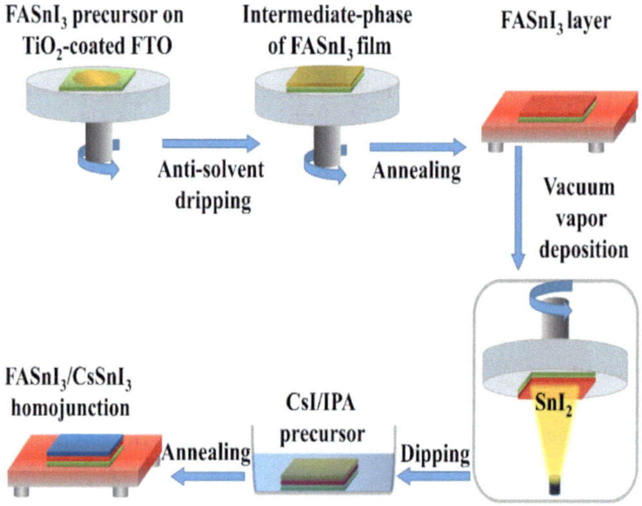

Figure 2.5. *LED homojunction.*

Source: Sajid, S., Alzahmi, S., Salem, I. B., Park, J., & Obaidat, I. M. (2023). Lead-free perovskite homojunction-based HTM-free perovskite solar cells: theoretical and experimental viewpoints. Nanomaterials, 13(6), 983. https://doi.org/10.3390/nano13060983, Creative Commons License.

2.4. POWER EFFICIENCY

Radiative power efficiency is multiplied by the luminous efficiency factor, which is wavelength dependent and peaks at 683 lumens/watt at 555 nm, to determine the luminous efficiency (Bhusal and Mascarenhas, 2012). GaN and AlGaInP LEDs are efficient than incandescent light sources in the red to green spectrum. LEDs surpass all other light sources in terms of efficiency if the external quantum efficiencies of III-V semiconductors are increased. For visible LEDs, semiconductor structures that are homojunction or heterojunction can be used (Figure 2.5). A heterojunction is created when narrow bandgap semiconductors are combined, enabling highly desirable characteristics that are not achievable in homostructures (Wang et al., 2014). Whereas two materials with distinct bandgaps are used in heterojunction LEDs, the p and n materials in homojunction LEDs are made of the same energy gap semiconductor. A homojunction and a diffused p/n junction constitute a standard GaAsP LED. Light emitted vertically from the junction is the only light that escapes because the GaAsP is formed on an absorbing GaAs substrate (Dong et al., 2016). For LED designers, this technique, known as bandgap engineering, (Figure 2.6) enables higher luminous efficiency than traditional homostructures.

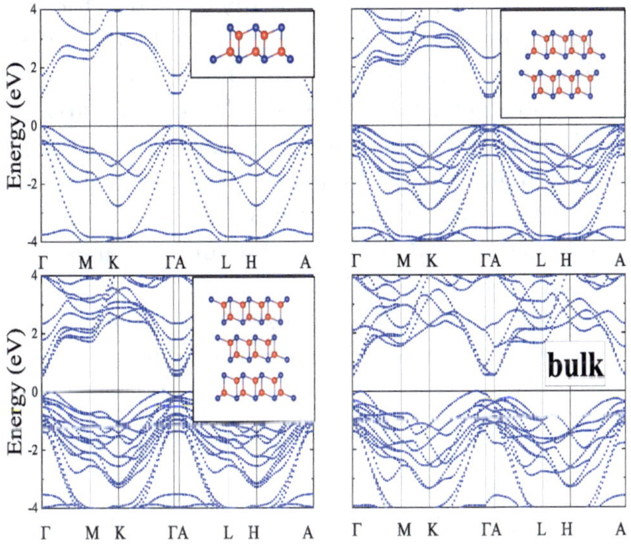

Figure 2.6. *Bandgap engineering.*

Source: Kuznetsov, V. G., Gavrikov, A. A., & Kolobov, A. V. (2023). Band gap engineering in ultimately thin slabs of CdTe with different layer stackings. Materials, 16(23), 7494. https://doi.org/10.3390/ma16237494, Creative Commons License.

The enhanced minority carrier injection efficiency and wide-gap material compatibility are the two main advantages of heterostructure LEDs. The wide-bandgap layers do not aid in photon reabsorption since they are transparent to the light produced in the narrow gap active layer. Due to the higher current injection efficiency at the heterojunction, heterostructure LEDs have a higher radiative efficiency than homostructure LEDs. Advanced devices are created by removing the minority carrier injection into the n-type material because the radiative efficiency of the p-side is typically higher than the n-side. A valence band discontinuity that serves as a hole potential barrier is produced by forming a heterostructure at the p/n interface. Confinement at the opposite side of the active layer rises with the addition of another heterostructure because the next heterointerface stops the injected electrons from diffusing out of the narrow-gap layer. A special type of heterojunction LED known as a quantum well (QW) LED occurs when the active region thickness is smaller than the electron de Broglie wavelength in semiconducting materials (Cheng, 2020). Since the electron and hole are wave-particles, they are tightly packed in space that distinct energy levels emerge. Emission is blue-shifted to higher energies than in bulk DH LEDs due to the recombination of electrons and holes in the subbands. The Group-III nitrides are especially attractive for dependable solid state device applications due to their many useful physical characteristics. The materials with a large bandgap have high thermal conductivity pathways and low dielectric constants. Group-III nitrides have relatively strong bond strengths and high melting temperatures (Costales et al., 2002). Compared to other II-VI and III-V materials, the high bond strengths may reduce dislocation motion and increase reliability. The nitrides are impervious to chemical etching, which might enable GaN-based devices to function in harsh conditions. These characteristics could result in a reliable device. The creation of device quality group-III nitride materials was made possible by a number of significant advances in the MOCVD growth of high quality GaN and InGaN (Fan et al., 2024).

2.5. SUBSTRATES

The lack of single crystalline GaN substrates or other single crystalline substrates with the same lattice characteristics as GaN is the primary cause of a number of issues with the epitaxial growth of nitrides. So, sapphire or SiC substrates have been used for the nitride epitaxial growth. The nitride epi-layer lattice mismatch with the substrate causes problems that must be resolved. The use of nucleation layers was one of the innovations in the development of device quality group-III nitride material. The quality of the GaN film grown at temperatures above 1000°C was found to be significantly improved by thin AlN or GaN nucleation layers formed using sapphire substrates at temperatures between 500 and 750°C (Md Taib et al., 2022). For SiC substrates, a thin AlN nucleation layer is typically deposited at high temperatures to start the growth process. GaN material of similar grade on both kinds of substrates might be produced in this method. The majority of GaN growth has been carried out on c-plane sapphire substrates. To find out if better structural matching can improve the GaN thin film performance, more substrate materials are presently being investigated. In addition to sapphire, a number of alternative substrates may provide superior lattice and thermal matching. Other substrate materials include ZnO, 3C-Sic, MgO, and 6H-Sic. ZnO provides a better structural match to the equilibrium wurtzite nitride because of its wurtzite structure, which has lattice constants of a = 3.32A, c = 5.213A (Nakamura and Chichibu, 2000). Compared to sapphire, 3C-SiC and MgO are cubic zinc-blende structures that have better structural and thermal matching to the nitrides. The cubic lattice constants for 3C-SiC and MgO are a = 4.36°A and a = 4.22°A, respectively. When employing non-equilibrium-based growth procedures, the nitrides can crystallize in a metastable zinc-blende structure, despite the fact that they are frequently found as the wurtzite (2H) polytype. Many of the challenges related to the deposition of device-quality material will be resolved by finding an appropriate substrate material that is thermally compatible and lattice matched with the GaN wurtzite structure (Figure 2.7).

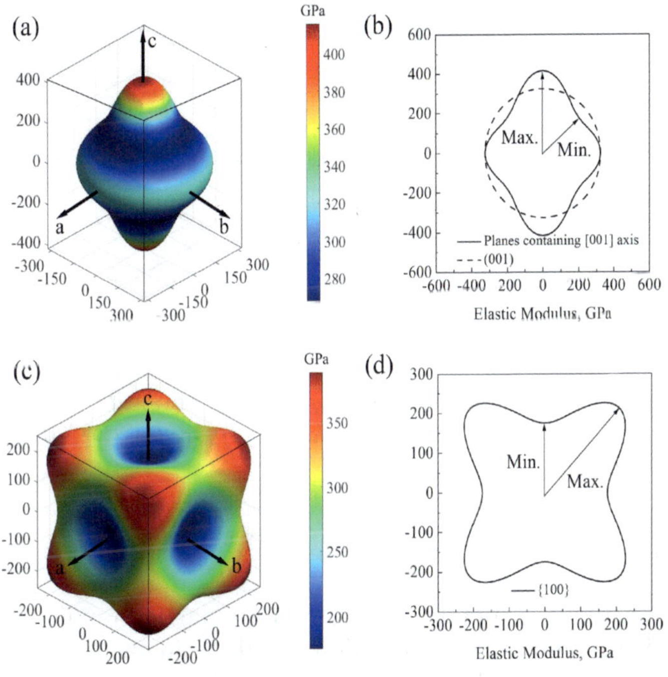

Figure 2.7. *GaN wurtzite (a-d).*

Source: Qin, H., Luan, X., Feng, C., Yang, D., & Zhang, G. (2017). Mechanical, thermodynamic and electronic properties of wurtzite and zinc-blende GaN crystals. Materials, 10(12), 1419. https://doi.org/10.3390/ma10121419, Creative Commons License.

One important finding in enhancing the surface shape and crystalline quality of GaN was the deposition of a low temperature buffer layer of AlN on sapphire substrates. Optimizing the buffer layer thickness can significantly improve the surface morphology of the resulting thick GaN layers (Hertkorn et al., 2007). Improved nitride films led to the observation of stimulated band-edge emission. The AlN buffer layer encourages GaN lateral growth and relieves the strain between GaN and sapphire. The buffer layer is highly dislocated, according to TEM data, while the dislocation density of following layers is significantly reduced.

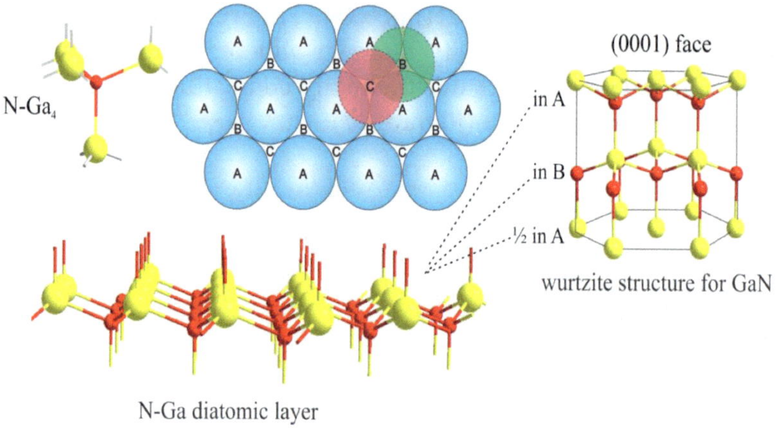

Figure 2.8. *GaN films.*

Source: Grodzicki, M. (2020). Properties of thin film-covered GaN(0001) surfaces. Materials Proceedings, 2(1), 30. https://doi.org/10.3390/CIWC2020-06833, Creative Commons License.

2.6. GAN

The nitridation of the sapphire substrate is crucial for regulating the defect structure and, therefore, the dislocation densities in overgrown GaN films, according to transmission electron microscopy (TEM), atomic force microscopy (AFM), and high-resolution X-ray diffractometry (XRD). The knowledge of X-ray rocking curves and their relationship to nucleation processes, and defect patterns in GaN films (Figure 2.8) on sapphire was explained based on these studies. The total dislocation density varies between 10^{10} and 10^{8} cm^{-2} when the sapphire surface nitridation time is altered prior to growth commencement (Ramesh et al., 2018). According to these structural analyses, GaN thin films have dislocation density of 1.7×10^{8} cm^{-2} (Datta et al., 2004). Using Mg with LEEBI treatment to create p-doping was one of the major advances in the development of GaN technology. Researchers found that Mg-doped GaN showed significantly better PL characteristics and significantly reduced resistivity under low electron beam irradiation (LEEBI) (Benzarti et al., 2023). The development of p/n GaN diodes with favorable characteristics aided this development. Researchers used high temperature thermal annealing in a nitrogen environment to generate significantly higher p-doping and homogeneous activation of Mg. For MOCVD material, post-growth treatment is necessary for the passivation in order to activate the dopants. Interstitial hydrogen is added during growth, and an H-Mg acceptor complex that passivates the acceptor is formed. An annealing step at a high temperature in an inert atmosphere can break this H-Mg bond. This shows that the main issue was the hydrogen compensation of Mg in the MOCVD development of GaN. The passivation of acceptors with hydrogen during CVD growth and the high activation energy of magnesium, the employed dopant (270 meV), complicate high room temperature p-doping (Wei et al., 2022). Dopant binding energies are influenced by the material effective mass and dielectric constant. Large binding energies are the result of the nitride system low dielectric constant and large effective masses. This is particularly obvious in p-type doping; the large hole mass causes the acceptor levels to be deep when comparing GaN to GaAs. High n-type doping has become challenging. Two factors contribute to this: the partial activation of the dopants at room temperature and the p-dopant being compensated by the high n-type background concentration. Current spreading issues and high contact resistances are caused by the low p-type doping. The production of LEDs and lasers with lower operating voltages and improved power efficiencies will be greatly aided by raising the p-doping level and creating new p-dopants. Silicon is the usual n-dopant, and n-type doping in GaN is standard. Just below the conduction band is the Si donor. Utilizing silicon as the donor can achieve controlled n-type doping. Silane (SiH_4) and disilane (Si_2H_4), which are normally diluted with hydrogen in the 200 ppm range, are the MOCVD precursors for n-type doping (El-Kareh and Hutter, 2024). The doping of GaN with silane readily reaches doping levels ranging from 1×10^{17} to $2 \times 10^{19} cm^{-3}$ (Nakarnura, 1997). Both atmospheric pressure MOCVD and low-pressure epitaxy have shown the ability to produce high-quality AlGaN films. Compared to TMGa, parasitic interactions between TMAI and NH_3 are likely to occur. It has been demonstrated that excessive prereactions in the gas phase can be effectively avoided by using two-flow-channel reactors with separate injection of the Group-III and Group-V precursors or reactor operation under low pressure.

2.7. ELECTRICAL AND OPTICAL PROPERTIES

For LEDs to have good electrical and optical properties, high-quality InGaN must grow. However, is harder to develop high-quality InGaN than GaN. Because of InN low dissociation temperature, InGaN growth must be carried out at significantly lower temperatures than GaN growth. Because of the high kinetic barrier for breaking nitrogen-hydrogen bonds, ammonia breakdown efficiency decreases with decreasing temperature. Due to extreme instability indium at typical GaN growth temperatures of over 1000°C, InGaN growth must be carried out at temperatures lower than 850°C (Thaler et al., 2010). However, indium droplet production was observed even on InGaN layers grown at temperatures lower than 800°C. Light-emitting diodes and laser diodes are examples of optoelectronic devices that depend on carrier injection into the active region. On opposite sides of the device, holes and electrons are inserted. They pass through several layers before coming together again in the active area. P-type doping, (Figure 2.9) or doping with acceptors, is necessary for the transport of holes through a layer; n-type doping, or doping with donors, is necessary for the transport of electrons.

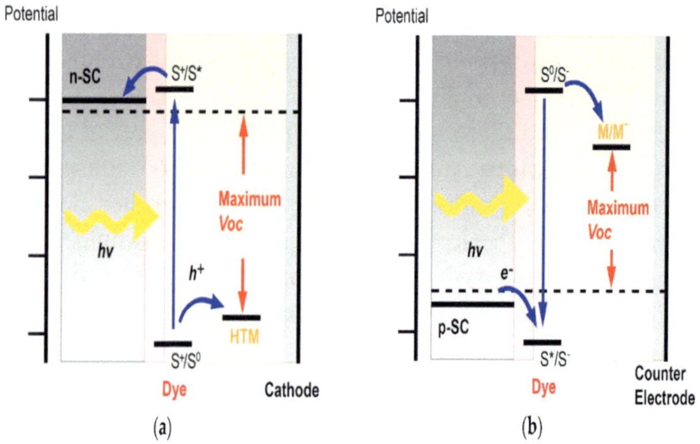

Figure 2.9. *P-type doping (a-b).*

Source: Li, M.-H., Yum, J.-H., Moon, S.-J., & Chen, P. (2016). Inorganic p-type semiconductors: Their applications and progress in dye-sensitized solar cells and perovskite solar cells. Energies, 9(5), 331. https://doi.org/10.3390/en9050331, Creative Commons License.

One of the important aspects of designing semiconductor devices is doping, which modifies a semiconductor conductivity. Wide-band-gap semiconductor doping present challenges, only one kind of doping is feasible. It has always been challenging to achieve p-type doping in the case of GaN. GaN as-grown frequently exhibits a tendency for n-type conductivity. Recent research has demonstrated that the n-type conductivity is caused by the insertion of impurities rather than native defects. Defects that are inherent to the crystal are known as native defects. Examples of these include antisites (GaN), self-interstitials (extra atom), and vacancies (missing atom). Deep levels are typically introduced into the semiconductor band gap by these defects. Sometimes the defect can function as a dopant and the defect-induced level is close to a band edge. However, natural defects typically serve as compensatory centers, they offset the dopant electrical activity. Doping levels must be managed and tuned to achieve the best possible device performance. Doping can be restricted by a number of factors, such as the dopant low solubility, high ionization energy, compensation from native defects, compensation from foreign impurities, and compensation from different dopant atomic configurations. A defect that only manifests at extremely low concentrations is not very problematic. Dopant impurities must be added in sufficient amounts to have the intended impact on conductivity. In theory, the formation energy of an impurity or native defect determines the concentration of the defect. Using pseudopotentials, a plane-wave basis set, and a supercell geometry, density-functional theory (DFT) in the local density approximation or LDA is the basis of the majority of theoretical studies that have been established for defects and impurities in GaN.

2.8. POTENTIAL

In the local density approximation (LDA), density function theory (DFT) enables the ground state of a many-body system to be described in terms of a one-electron equation with an effective potential (Yabana et al., 2006). A Hartree potential, an ionic potential resulting from the atomic cores, and exchange and correlation potential that characterizes the many-body characteristics make up the total potential. For many different solid-state issues, this method is useful. The valence electrons of molecules and solids control the majority of their properties; the core electrons may typically be eliminated by using a pseudopotential to represent the ionic core, which is made up of the nucleus and inner electron shells. There are two difficulties with using pseudopotentials in calculations for nitride semiconductors. Nitrogen has a deep potential that necessitates a significant number of terms in the Fourier expansion, a planewave basis set that is frequently used to expand potentials and wave functions in a variational method. The 4d states of In and the 3d states of Ga cannot be regarded as core states. These energy positions are quite near to those of the N 2s states, and if the d states are regarded as a component of the core, the connection between these levels might not be sufficiently explained. Because the d electron wave functions are more confined than those of the s and p states, including them as valence states significantly increases the computational load. Accurate findings can occasionally only be obtained by including the d states as valence states. However, in many instances, a correction to the exchange and correlation potential, often referred to as the nonlinear core correction can sufficiently account for the effect of the d states. Nitrogen vacancies raise the electron concentration when they are intentionally produced, for as during irradiation or ion implantation. They cannot, however, be the cause of n-type conductivity because of their high formation energy, which makes it extremely unlikely that nitrogen vacancies would arise on their own during the growth of undoped or n-type GaN. Unintended n-type conductivity in GaN must therefore be attributed to other factors, most likely the addition of donor impurities like silicon or oxygen. Nitrogen vacancies in p-type GaN have a low formation energy, which makes them a reasonable compensating center when acceptor doping is present. It has been suggested that the commonly observed yellow luminescence (IT) originates from the gallium vacancy. The GaN YL is a broad band of luminescence with a center of 2.2 eV (Reshchikov, 2023). Because the gallium vacancy is an acceptor-type defect, when the Fermi level rises, its formation energy falls. n-type GaN is more likely to have gallium vacancies than p-type GaN. Experimental findings showing YL suppression in p-type material are related with this tendency. The intensity of the YL rises with increasing n-type doping. In line with the gallium-vacancy concept, researchers also discovered that the defects causing the YL function as compensatory centers. Gallium vacancies are more prevalent in Ga-poor materials and less prevalent in Ga-rich materials. Higher trimethyl-gallium (TMGa) flow rates in samples generated by metal-organic chemical vapor deposition (MOCVD) have been shown to experimentally reduce the YL. The Ga vacancy has a deep level that is roughly 1.1 eV above the valence band. So, transitions between this deep level and the conduction band show the accurate energy to account for the YL. The YL is caused by a transition between a deep acceptor level and a shallow donor level around 1 eV above the valence band (Soh et al., 2004).

2.9. IMPURITIES

In GaN, gallium vacancies can combine with donor impurities, especially oxygen, to form complexes. This compound exhibits a deep level again around 1.1 eV above the valence band due to its structure being extremely close to that of the gallium vacancy. Because the complex formation energy is lower than that of the isolated vacancy, oxygen can increase the concentration of Ga vacancies and, therefore, the YL. The rise in YL intensity in the vicinity of the interface with the sapphire substrate, where the oxygen content is also known to be higher, can therefore be explained by complexing with oxygen. Positron annihilation measurements provide the evidence of a relationship between Ga vacancies and YL. Ga vacancies were present in both GaN bulk and epitaxial layers at concentrations ranging from 10^{17} to 10^{18} cm^{-3} (W. Jiang et al., 2017).

The nitrides have a tendency to be n-type conductive. The spontaneous creation of nitrogen vacancies was the cause of this. The nitrogen vacancies are rare to ascend in n-type material. Therefore, the addition of dopant impurities must be responsible for the observed n-type conductivity. In 1983, it was suggested that oxygen might be the source of n-type conductivity in GaN (Pearton et al., 1999). The impact of oxygen on the electrical and optical characteristics of GaN produced by MOCVD was observed by Chung and Gershenzon in 1992 (Tansley et al., 1997). Water was injected to introduce oxygen. In contrast to nitrogen vacancies as the source of the conductivity, their carrier concentrations dropped as the growth temperature increased. At the higher growth temperatures, less oxygen was being assimilated. It was challenging to disprove the that nitrogen vacancies were the cause of the n-type behavior. Unintentionally doped n-type GaN samples had silicon or oxygen concentrations high enough to account for the electron concentrations. After electrically characterizing both intentionally and accidentally doped samples, researchers came to the conclusion that silicon was responsible for the n-type conductivity in the latter. Researchers discovered evidence of a second shallow donor, which was identified as oxygen, having a higher activation energy.

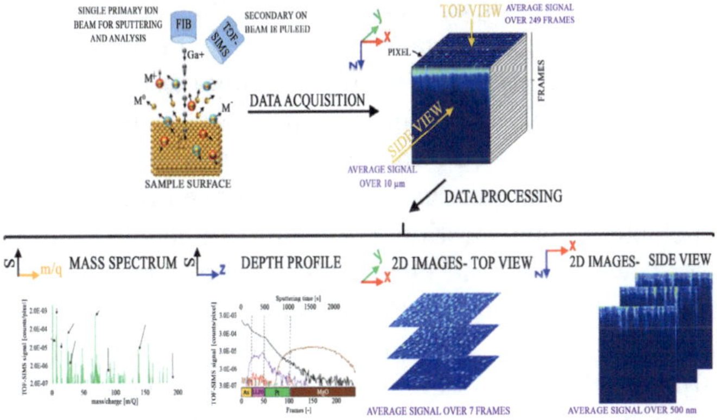

Figure 2.10. *Secondary-ion mass spectroscopy (SIMS).*

Source: Priebe, A., & Michler, J. (2023). Review of recent advances in gas-assisted focused ion beam time-of-flight secondary ion mass spectrometry (FIB-TOF-SIMS). Materials, 16(5), 2090. https://doi.org/10.3390/ma16052090, Creative Commons License.

When applied to hydride vapor phase epitaxy (HVPE) material, secondary-ion mass spectroscopy (SIMS) (Figure 2.10) also reveals silicon or oxygen levels that match the electron concentration as established by electrical measurements (Hoffmann, 1996). GaN bulk crystals produced at high temperatures and high pressures have been shown to exhibit high levels of n-type conductivity. Recent high-pressure investigations have demonstrated that these sample properties closely resemble those of intentionally oxygen-doped epitaxial films. Accidental oxygen integration is responsible for bulk GaN n-type conductivity. There are many sources of contaminating impurities, but the most common source in MOCVD is water, which is the primary NH_3 impurity. In the case of InGaN growth, even high-purity NH_3 can be a substantial source of oxygen contamination. Water may be eliminated by purification. Researchers investigated the contamination of GaN epitaxial films by contaminants originating from the substrate. Large concentrations of Si were observed in films formed on SiC, ZnO was incorporated, and O was incorporated when films grew on sapphire. There may have been impurity diffusion at the growth temperature since the inclusion took place in an area larger than the defective near-substrate interface layer. Based on Raman and luminescence studies, researchers also documented impurity diffusion into the near-substrate region of HVPE-grown layers. Oxygen was most likely the contaminant, leaking from a ZnO buffer or the sapphire substrate. Researchers examined the addition of different impurities to GaN produced via reactive MBE using SIMS and Hall effect measurements (Tingberg et al., 2017). The unintended n-type conductivity was caused by the nitrogen vacancy. Specific collection of samples they utilized in their investigation is not ideal for ruling out oxygen contamination. Their results are completely compatible with oxygen being the donor because all of their samples actually had enough oxygen to account for the electron concentration. However, an increase in oxygen concentration does not result in an increase in electron concentration. Oxygen integrates in other ways besides on nitrogen sites, where it is an electrically active donor. This extra oxygen could reflect as Ga_2O_3 precipitates. Scientists have identified some drawbacks when attempting to derive dopant concentration and ionization energy information from Hall-effect observations. Research on unintentionally doped HVPE-grown layers revealed the existence of a thin, extremely conductive layer close to the sapphire substrate interface. This layer functions as a parallel conduction channel. This layer's high conductivity might be caused by oxygen contamination. For the purpose of intentional n-type doping of GaN, AlGaN, and InGaN, silicon is utilized. Silicon doping is accomplished by employing a solid Si source in MBE. Since silicon has a relatively low formation energy, it is integrated in GaN. Ion implantation doping is quite challenging in nitrides. Only at temperatures higher than those at which GaN starts to break down can the damage caused by the implantation process be annealed. Silicon implanted at a dose of 10^{16} cm^{-2} and annealed at 1100°C could achieve up to 50% activation (Sameshima et al., 2018). Only at temperatures greater than normal growth temperatures does the damage appear to be effectively annealed. Encapsulation or operation at high nitrogen pressure can stop the film from decomposing. Researchers used encapsulation with AlN films and fast heat processing at 1400–1500°C to produce Si implant activation efficiencies over 90% (Cao et al., 1998). The activation energy of silicon in GaN is 17 meV. The ionization energy is sensitive to the dopant concentration, as was observed in the case of oxygen. The DX transition is absent in silicon donors. Silicon was shown to continue to be a shallow donor in $Al_XGa_{1-x}N$ alloys up to x = 0.44 and in GaN under pressure up to 25 GPa. Si will continue to be a shallow donor over the composition range i.e., up to x = 1. Silicon doping of InGaN well and barrier layers has been shown to enhance the device properties of laser diodes. Better structural characteristics and screening of the polarization fields, which raises emission efficiencies, are possible explanations for the impact.

2.10. MOBILITY

The mobility of the carriers as well as the carrier concentration determine a doped layer resistivity.

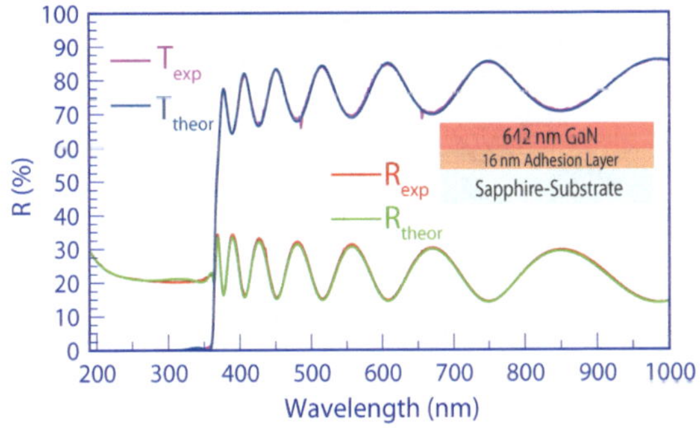

Figure 2.11. *GAN on sapphire.*

Source: https://upload.wikimedia.org/wikipedia/commons/a/ae/GaN_on_Sapphire_Reflectance.png, Creative Commons License.

The existence of defects in epitaxial layers complicates the analysis of mobilities; it is challenging to determine how those defects affect carrier mobilities. At normal temperature, high-quality films typically have electron mobility values between 500 and 1000 cm^2/Vs and hole mobility between 5 and 10 cm^2/Vs (Kuhl, 1993). Using a model created for polycrystalline materials, some researchers attributed the observed behavior to possible obstacles brought on by electron traps at grain boundaries. Columnar grains that are twisted relative to one another are produced by epitaxial development of GaN on sapphire, (Figure

2.11) and it is probable that there would be a significant defect density along the boundaries between such regions. Potential barriers created by band bending close to the boundaries have an impact on electron mobility. The process of modulation doping, which is used on heterostructures, involves introducing dopants into a material with a wider band gap so that the generated carriers can move into a lower band gap, undoped material.

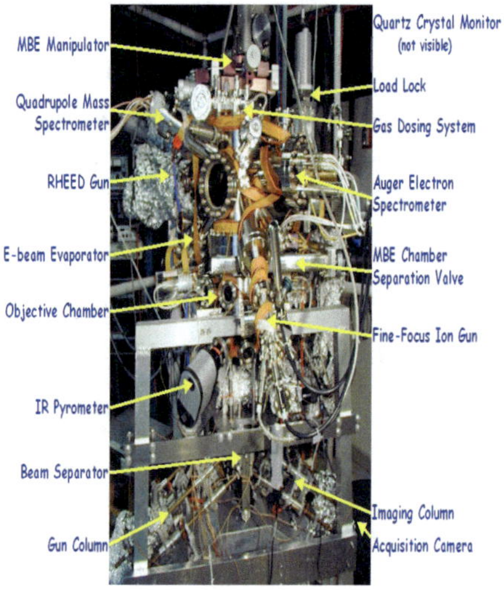

Figure 2.12. *LEEBI.*

Source: https://upload.wikimedia.org/wikipedia/commons/c/c8/UIUC_LEEM.jpg, Creative Commons License.

The primary benefit is that ionized impurities do not cause scattering for carriers traveling in the lower-gap, undoped material. Modulation doping of strained-layer superlattices (SLS) for n-type and p-type cladding layers in laser diodes was described by Nakamura et al. in 1998a. The 2.5-nm-thick undoped GaN and 2.5-nm-thick Si-doped $Al_{0.4}Ga_{0.86}N$ layers made up the n-type layer. Although the $Al_{0.4}Ga_{0.86}N$ was doped with magnesium, the p-type SLS was similar. There have been reports of improvements in device performance. Only the GaN layers in the GaN/AlGaN superlattice cladding layers were doped. Therefore, rather than modulation doping, it appears that structural enhancements account for the majority of the superlattice cladding layer advantages. Magnesium was used as the dopant in the first successful p-type doping of GaN. They discovered that although the Mg-doped MOCVD layers were not p-type as-grown, they were electrically active after an activation procedure. To accomplish this activation, researchers employed low-energy electron-beam irradiation, or

LEEBI (Figure 2.12). They demonstrated how MOCVD-grown magnesium-doped films, which are initially very resistive, may undergo thermal annealing in a N_2 environment to become p-type conductive. After annealing their samples in NH_3 at 600°C, Nakamura et al. discovered that the resistivity had considerably increased (Miyajima et al., 2023). They came to the conclusion that the passivation of acceptors was caused by atomic hydrogen created by NH_3 dissociation. A comprehensive study of the behavior of hydrogen in GaN, both as an isolated impurity and in its interactions with dopant atoms, is essential due to the critical role that hydrogen plays in p-type doping. The preferred p-type dopant is magnesium. Bis-cyclopentadienyl magnesium is commonly added by flowing during MOCVD development or from a solid Mg source in MBE. The maximum room-temperature hole concentration for magnesium doping that has been documented is $3 \times 10^{18}cm^{-3}$, together with a resistivity of 0.2 cm and a mobility of 9 cm^2/Vs. It is quite difficult to get both a high mobility and a high hole concentration. By creating a defect band where acceptor ionization is not required, one may be able to achieve hole concentrations beyond $10^{18}cm^{-3}$, but at the expense of limited mobility. Higher resistance follows from the reduced mobility. Using the variable-temperature Hall effect, the Mg activation energy was determined to be 208 meV for low Mg concentrations (Zhao et al., 2014). It was discovered that consideration of the Coulomb interaction between ionized acceptors was crucial. As the degree of ionization increases, this interaction results in decreased activation energies.

2.11. HOLE CONCENTRATION

With magnesium as the dopant, it has been challenging to get high hole concentrations because, although the Mg concentration can be elevated, the hole concentration eventually drops. The significant limiting factor is magnesium solubility. Competition between the production of Mg_3N_2 and the inclusion of Mg acceptors determines the solubility (Malin et al., 2021). The substitutional Ga site is preferred by magnesium. Limited doping efficiency has occasionally been attributed to the incorporation of magnesium on interstitial sites (Mg) or substitutional nitrogen sites (MgN); however, the estimated formation energy of Mg on these sites is extremely high. The observed limitations in doping cannot be explained by the adverse integration of magnesium on these locations. The characteristics of GaN are strongly affected by hydrogen. Many growth methods involve the developing material to high hydrogen concentrations, including hydride vapor phase epitaxy (HVPE) and metal-organic chemical vapor deposition (MOCVD) (Tassev and Vangala, 2019). The presence of hydrogen has significant effects on the material p-type doping because it causes passivation of the acceptors during growth, necessitating a post-growth processing step to make the acceptors electrically active. However, the presence of hydrogen during development is advantageous, as it increases acceptor solubility and inhibits compensation. Important details regarding the interaction of hydrogen with impurities can be collected from the behavior of isolated interstitial hydrogen. Hydrogen-donor compounds are uncommon in n-type GaN due to the low solubility and diffusivity of hydrogen. In p-type GaN, hydrogen is found in the positive charge state behind a nitrogen atom in an antibonding site. It also assumes the antibonding configuration and is electrostatically drawn to negatively charged acceptors. This indicates that H is not next to the Mg atom in the case of Mg acceptors, but rather binds to a neighboring N atom. This is a unique structure for an acceptor-hydrogen complex; in other semiconductors, when hydrogen binds to acceptors, it takes on the bond-center position. The complex vibrational frequency is indicative of a N-H bond than a Mg-H bond. In the harmonic approximation, the computed vibrational frequency is 3360 cm^{-1}. This computed frequency may be lowered by anharmonic effects. The determination of the local vibrational modes (LVM) of the Mg-H complex in GaN has yielded valuable information about the complex structure as well as confirmation of the importance of hydrogen in GaN.

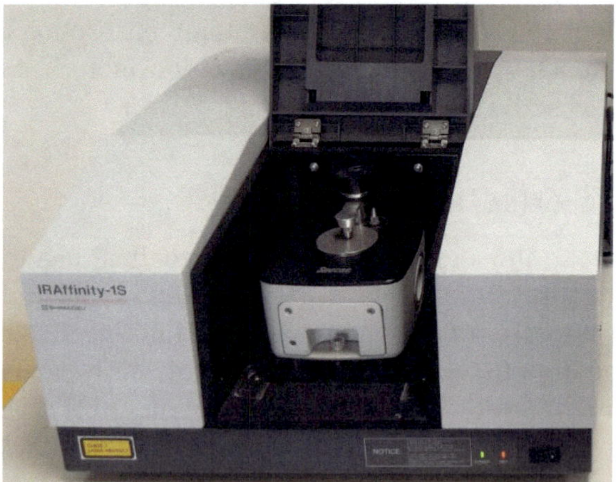

Figure 2.13. *Fourier-transform infrared absorption spectroscopy.*

Source: https://upload.wikimedia.org/wikipedia/commons/1/1b/FTIR_Spectrometer_%2B_ATR.jpg, Creative Commons License.

Mg-doped GaN produced by MOCVD was subjected to Fourier-transform infrared absorption spectroscopy (Figure 2.13) (Sun et al., 2001). The as-grown sample showed an LVM at 3125 cm^{-1} and was semi-insulating. Following thermal activation, the sample developed p-type conductivity and the LVM intensity decreased. After two hours of exposure to monatomic deuterium at 600°C, the resistivity increased and a new absorption line at 2321 cm^{-1} appeared. Following a thermal activation treatment, this line disappeared. The presence of hydrogen in the compound was demonstrated by the isotopic shift. It is established that the Mg-H complex has a strong N-H bond when the LVM is assigned to the stretch mode of the complex. The spectroscopic evidence proves that hydrogen and a shallow acceptor form a complex, and that the dissociation of Mg-H complexes is what causes the acceptor activation process. The heat breakdown of Mg-H complexes is compatible with the decrease in LVM intensity upon post-growth activation of the p-type conductivity (Popovici and Morkog, 2000). The Mg-H complex predicted binding energy is 0.7 eV. Although hydrogen is present in the Mg-doped layer, this result suggests that hydrogen is not bound to the acceptors at the growth temperature. Hydrogen atoms disperse quickly and eventually form hydrogen-acceptor complexes. Since the hydrogen must pass through a barrier in order to be extracted from the Mg atom, the dissociation energy of the complex, which was estimated to be around 1.5 eV, is greater than the binding energy. However, because the production energy of H^+ is less than that of V_N, compensation by nitrogen vacancies is inhibited when hydrogen is present. Rather than nitrogen vacancies, hydrogen is preferentially incorporated as a compensatory center. The Fenni level equilibrium point is shifted higher in the gap because the formation energy of hydrogen is lower than that of V_N. This results in a lower formation energy and, so, a higher concentration of magnesium. Therefore, adding hydrogen has advantages: it suppresses native defects and increases the concentration of the acceptor. It may be possible to maximize acceptor doping by adjusting the atomic hydrogen concentration during development. The drawback of including hydrogen is that the acceptors are fully compensated. At higher temperatures, the complex should be able to dissociate due to the height of the dissociation barrier. However, it has been revealed experimentally that Mg-doped MOCVD-grown GaN requires temperatures higher than 600°C to activate (Wei et al., 2023). This suggests that the activation mechanism involves more than only the dissociation of Mg-H complexes. As long as it stays in the p-type layer, hydrogen can still compensate the acceptor after leaving the magnesium acceptor since it still acts as a donor. The hydrogen must be eliminated from the p-type layer; alternatively, it can be neutralized, for example, by attaching itself to defects. GaN high H_2 production energy prevents the creation of H_2 molecules, which is normal in other semiconductors following acceptor passivation. The nitrogen vacancy functions as a compensation center and has a substantially lower formation energy in p-type material than it does in n-type GaN. The p-type materials can develop a specific concentration of nitrogen vacancies, which serve as compensation centers.

2.12. Interactions

The interactions between hydrogen and nitrogen vacancies may be the cause of the formation and disappearance of photoluminescence (PL) (Figure 2.14) lines during postgrowth annealing of Mg-doped layers produced by MOCVD (Wang et al., 2021). Growth can result in the formation of complexes between hydrogen and nitrogen vacancies; the V_NH^{2+} complex predicted binding energy, expressed with respect to interstitial H in the positive charge state, is 1.56 eV (Kumar and Krishnan, 2021).

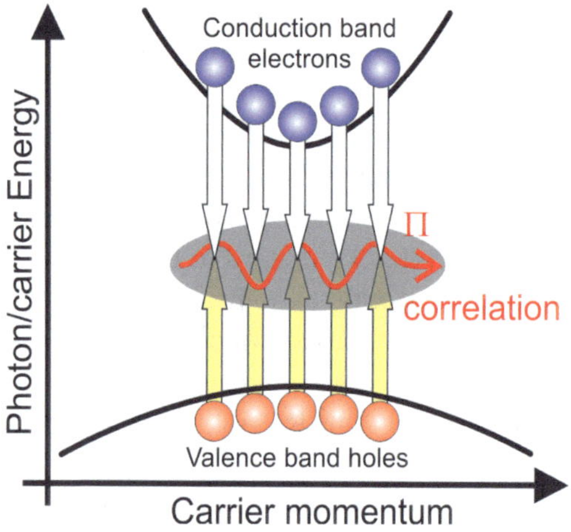

Figure 2.14. *Photoluminescence.*

Source: https://upload.wikimedia.org/wikipedia/commons/b/b8/Photoluminescence_schema.png, Creative Commons License.

The behavior of PL lines during annealing for acceptor activation may be explained by the dissociation of this complex, which results in isolated nitrogen vacancies. The equilibrium concentration of the impurity in the lattice, which is established by the formation energy, is equivalent to the solubility of a substitutional acceptor. The probability of inclusion on sites other than the substitutional site is determined by the formation energy of the impurity in other configurations. Compared to magnesium, the computed ionization energies for zinc and calcium are higher. Be is as soluble as magnesium. Although its estimated ionization energy is marginally lower than that of magnesium, the difference is not statistically significant because it falls within the computation error. Be may experience problems with inclusion on interstitial sites because to its small size. Researchers calculated an ionization energy of 0.34 eV for Zn using photoluminescence (PL) (Zatsepin et al., 2016). Since the Zn acceptor has a high ionization energy, doping with Zn has not produced p-type conductivity. For double-heterostructure blue LEDs, zinc was employed as a dopant in InGaN. Around 800°C was used to generate InGaN layers, with diethylzinc serving as the Zn supply. Zn giving rise to a relatively deep acceptor level is consistent with the Zn doped films producing strong luminescence at energy 0.5 eV lower than the emission from Si-doped InGaN. Scientists used dimethyl-cadmium and growth temperatures of about 800°C to study Cd doping of InGaN. As long as oxygen is present as a co-dopant, some researchers found high levels of p-type doping employing Be in MBE; however, the experimental results regarding Be are not definitive. Because of the lattice hardness and poor diffusivity, implantation damage is very difficult to eliminate by annealing in GaN. It is therefore challenging to accurately evaluate the characteristics of the implanted acceptors when implantation defects are present, and no definitive findings have been made. The majority of optoelectronic devices require heterojunctions in order to provide both optical and carrier confinement. GaN, AlN, InN, or their alloys form heterojunctions in the nitride-based devices. The band discontinuities in the valence and conduction bands are the crucial characteristics that define such heterojunctions. At nitride heterostructures, strain has a significant impact on the band offsets and band structure. AlN and GaN have a lattice mismatch of roughly 3%, while GaN and InN have a mismatch of more than 11% (Wang and Yoshikawa, 2004). Thin semiconductor layers can grow on top of a layer with a differing lattice constant due to growth techniques used to fabricate nitride devices, primarily MOCVD and MBE. A pseudomorphic interface forms and the overlayer is uniformly stretched to match the substrate in-plane lattice constant when its thickness is smaller than that of the critical layer. This crucial layer thickness has been shown to be quite substantial for InGaN growth on GaN. The lattice mismatch between the epitaxial layer and the substrate, or between the epitaxial layer and the layer it is grown on, determines the amount of strain.

CHAPTER 3

PHOTONICS IN SEMICONDUCTORS LASER

Contents

3.1. INTRODUCTION

When the Bernard-Duraffourg condition is met by appropriate carrier injection, a semiconductor that is far from thermodynamic equilibrium can exhibit a net gain (Gunnarsson et al., 2023). This condition must be met to create a carrier inversion. A p-n junction made of degenerate semiconductors can be electrically pumped so that the quasi-Fermi levels fall in the valence and conduction bands, respectively. In 1962, the first report of laser action in a semiconductor was made utilizing p-n homojunctions (Panish and Hayashi, 1974). Homojunctions are no longer employed in light-emitting diodes because of several disadvantages; because of the large threshold current density at normal temperature (105A/cm^2), they can only operate CW at cryogenic temperatures (77 K). Due to their advantages, double heterostructures significantly lower the threshold current density at room temperature (103A/cm^2) (Srivastava, 1990). Similar to LEDs, the active region is positioned between the n-type and p-type cladding layers with a huge energy gap. It typically has a thickness of 0.1–0.2 μm. Outside of the active zone, the photons are not absorbed. Due to the differing energy gaps between the active region and the cladding layers, the injected electrons and holes are effectively contained inside the active layer, increasing their concentrations at a given current density. A waveguiding structure is created as a result of the active layer having a higher refractive index than that of the p and n-type cladding layers (Xu et al., 2020). This effectively confines the laser beam within the active layer. The thickness and refractive indices of the semiconductors determine how many modes are present in this waveguide. A single transverse mode can be achieved with the correct design.

3.2. FEEDBACK

Suitable positive optical feedback is necessary to achieve laser activity. The necessary feedback for laser activity is provided by Fresnel reflection at the semiconductor – air interfaces (Brown and Arnold, 2010). Cleavage along crystal planes induces the parallel end faces. This is caused by the high gain that can be achieved from semiconductors as well as their high refractive index. In order to determine the output power and the threshold condition for laser activity, we can develop the rate equations of a semiconductor laser. Number of devices have been implemented to improve laser performance. It is known that the transverse mode profile is uniform. N is the carrier density in the rate equations. The rate equation for N accounts for the recombination rate per unit volume in the active zone as well as the electron generation rate from electron injection.

$$R_P = \eta_i \frac{I}{eV}$$

When 'I' denotes the pumping current and ηi denotes the injection efficiency, which yields the proportion of injected current producing carriers in the active zone. This yields the generation rate per unit volume, or pumping rate. All radiative and nonradiative recombination mechanisms are connected to the recombination rate. The rate at which the photon density, φ, in the laser cavity changes is provided by the rate equation. The spontaneous emission rate (R_{sp}, T) and the net stimulated recombination rate (R_{st}) are correlated with the photon generating rate. Only a portion of the total spontaneous emission, $\beta_{sp}R_{sp}$,T, which is known as the spontaneous emission factor, is linked into the lasing mode because the resonant cavity oscillates on a single mode. Within the volume, V, of the laser active region, generation activities that are related to laser activity must be taken into account. The photons occupy a larger volume (V_p) than electrons do in the active region. Because of this, it is useful to include a confinement factor, which shows the percentage of laser beam strength in the active area. The rate of photon density generation can be expressed as follows:

$$\Gamma = \frac{v}{v_p}$$

3.3. CAVITY

In a laser cavity, photon losses are due to mirror reflectivity, which is always less than 1, absorption, and scattering in some optical components within the cavity. It is helpful to introduce the photon lifetime, τ_c, which takes into account the reduction of photons in the cavity due to cavity losses, in order to measure the rate of photon density decay in a particular cavity mode. For example, $T_0 \geq 120$ K at room temperature in the GaAs/AlGaAs double-heterostructure laser, T_0 is between 150 and 180 K for GaAs/AlGaAs quantum well lasers and greater than 200 K for strained InGaAs/AlGaAs quantum well lasers (Figure 3.1) (Gribkovskii, 1995). Another benefit of quantum well devices is the rise in characteristic temperature. T_0 is between 50 and 70 K at ambient temperature for InGaAsP/InP double-heterostructure and quantum well lasers that emit between 1.3 and 1.55 μm (Dutta, 2002). The performance of these lasers at high temperatures is limited by these relatively low values, which are caused by Auger recombination and leakage current. The optical properties of quantum wells, such as their larger differential gain and weaker dependence of this gain on temperature, as well as the potential to use strained layers to improve the optical characteristics and the band gap tunability, make them suitable for use as active materials in diode lasers. A large threshold current results from the small confinement factor caused by the thin active materials. Separate confinement heterostructures (SCHs) with different configurations are employed to increase the confinement factor (Tsang, 1987). The barrier layers, which have a thickness of about 0.1 μm and a larger refractive index than the cladding layers, are responsible for achieving the beam confinement. While the quantum well confines the carriers, it makes a very less contribution to beam confinement. The refractive index of the inner confinement structure gradually transitions from a maximum value at the center of the structure to the value in the cladding region in the case of the graded-index SCH. A single quantum well or a multiple quantum well, which is made up of several alternating thin layers of semiconductors with wide and narrow energy gaps, can serve as the active material. Because of the higher driving current and low output beam quality, the simple laser structure is not suitable, even in the case of a double heterostructure arrangement.

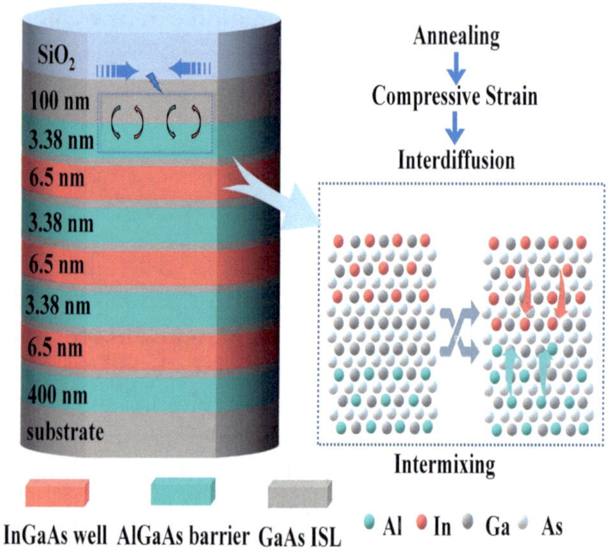

Figure 3.1. *InGaAs/AlGaAs quantum well lasers.*

Source: Yang, Z., Zhang, S., Ma, S., Shi, Y., Liu, Q., Hao, X., Shang, L., Han, B., Qiu, B., & Xu, B. (2023). Effects of thermal-strain-induced atomic intermixing on the interfacial and photoluminescence properties of InGaAs/AlGaAs multiple quantum wells. Materials, 16(17), 6068. https://doi.org/10.3390/ma16176068, Creative Commons License.

3.4. LASER OPERATION

Stable laser operation and high-quality output beams depend on the lateral confinement of carriers and photons. In order to prevent leakage currents, which could circumvent the active zone and significantly lower the laser efficiency, the injected carriers must be confined laterally. Gain-guided structures and index-guided structures are the two primary classes into which a variety of laser configurations fall (Agrawal and Dutta, 1993). The current flow is significantly decreased by the addition of an oxide layer, which forces the current injected from the positive electrode to flow in a stripe with a small or narrow width (5 μm). The beam transverse dimension is weakly confined in the direction parallel to the junction due to a lateral waveguiding. Since the optical beam is guided by the shifting of the gain, the structure is known as gain-guided. The optical beam is confined to the basic mode in the direction perpendicular to the junction by the double heterostructure or by the separate confinement structure, whereas the laser mode is confined to the transverse mode in the direction parallel to the junction if the stripe width is less than 10 μm (Eliseev, 1980). The substantial absorption of the unpumped active layer components, which results in losses, is a significant drawback of the gain-guided structure. Because of the relatively limited lateral waveguiding effect, even slight changes in the refractive index can cause unstable operation. Because of the increased spontaneous emission, the output beam optical spectrum is multimodal. A small stripe of the active ingredient is incorporated into semiconductor layers that have a lower refractive index and a wider band gap. Lateral single-mode functioning is made feasible by the active layer decreased lateral width. A reverse-biased diode is created by the n-type burying layer and the p-type current blocking layers, preventing the current from avoiding the active region. Strong index-guided laser structures can be used to produce steady laser performances with low threshold currents (<10 mA) and lateral single-mode operation (Botez and Connolly, 1981). These factors account for their extensive use in commercial products, even though their structure necessitates intricate fabrication processes. A diode laser output spectrum is determined by the active medium optical gain curve and the optical resonator configuration. The length controls how far apart the longitudinal modes are in a Fabry–Perot (FP) cavity. While many FP cavity resonances are detectable throughout a broad spectrum in a typical semiconductor laser below threshold, only a few modes near the gain curve peak oscillate above threshold. The formula for the frequency difference between adjacent modes is:

$$\Delta v = \frac{C}{2Ln_g},$$

Reducing the cavity length 'L' until the mode spacing is greater than the gain curve width is one method of achieving single-mode operation, cavity lengths of a few micrometers are needed for this. Vertical cavity surface-emitting lasers are the devices that resemble to these short cavities, which can be created along the epitaxial deposition axis (Rantamäki and Guina, 2021).

3.5. SINGLE MODE

The utilization of distributed Bragg reflector lasers (Figure 3.2) is another method for single-mode laser operation. Because of an optical fiber chromatic dispersion, single-mode laser operation is crucial for optical-fiber communications. The output beam is always diffraction limited in the plane perpendicular to the junction because of the small beam size in the direction perpendicular to the junction.

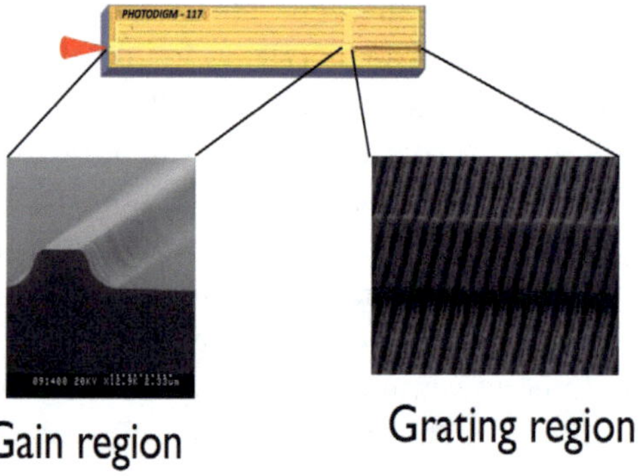

Figure 3.2. *Distributed Bragg reflector lasers.*

Source: https://upload.wikimedia.org/wikipedia/commons/0/0b/DBR_laser_schematic.jpg, Creative Commons License.

The beam is diffraction-limited in the plane parallel to the connection if the stripe width is less than about 10 μm. The beam ellipticity rotates by 90° at the laser cavity output, assuming an elliptical beam on the exit face of a laser with dimensions of 1 μm × 5 μm. This is because the divergence is greater in the direction perpendicular to the junction. Appropriate optical devices can be employed to correct for beam astigmatism when a circular beam is needed. It has been demonstrated that quantum confinement effects in quantum well structures can lead to enhancements in optical characteristics, which are crucial for laser applications. The primary distinction between quantum wells and bulk materials, which essentially dictates their optical and electrical characteristics, is the form of the density of states. Quantum wires and quantum dots (QDs) (Figure 3.3) are the result of the progression of quantum confinement in two and three dimensions (Zhu et al., 2007). All three dimensions of carrier motion in QDs are constrained. QDs display distinct atomic-like states, the energy of which increases as the QD size decreases and is dependent on the dot size. The emission wavelength and the associated energy gap are size dependent and can be adjusted across a wide energy range by employing QDs of varying sizes made of II-VI, III-V, and IV-VI semiconductors (de Melo, 2023). For technical applications, this broad spectrum tunability is crucial, especially for the creation of lasers that emit in spectral areas that are difficult for conventional semiconductor lasers.

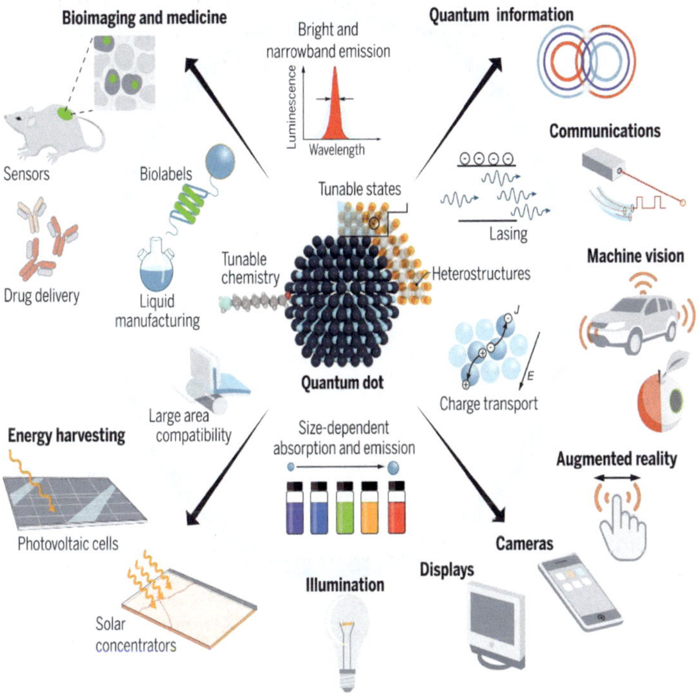

Figure 3.3. *Quantum dots.*

Source: García De Arquer, F. P., Talapin, D. V., Klimov, V. I., Arakawa, Y., Bayer, M., & Sargent, E. H. (2021). Semiconductor quantum dots: Technological progress and future challenges. Science, 373(6555). https://doi.org/10.1126/science.aaz8541, Creative Commons License.

The wider distance between the atomic-like energy levels in QDs prevents thermal depopulation of the band edge states, which lowers the threshold current temperature dependency and the optical gain threshold for laser operation. Researchers first suggested using QDs as an active component in a laser in 1982. QDs were referred to as three-dimensional quantum wells. Because quantum boxes have a significantly different density of states than bulk and quantum wells, they theoretically examined the benefits of quantum boxes in III-V material systems. These benefits included reduced threshold current density, reduced total threshold current, improved differential gain, and high spectral purity. With a low threshold current density (120 A/cm^2) at 77 K and a characteristic temperature T_0 = 350 K, the first QD laser demonstration was made. Growth techniques and laser designs were quickly enhanced, and by 2000, QD lasers were outperforming their quantum well counterparts. The creation of a three-dimensional structure during the growth of InAs/GaAs-strained quantum wells by molecular beam epitaxy (MBE) (Figure 3.4) was documented by Goldstein in 1985 (Sinusia and Gómez, 2023).

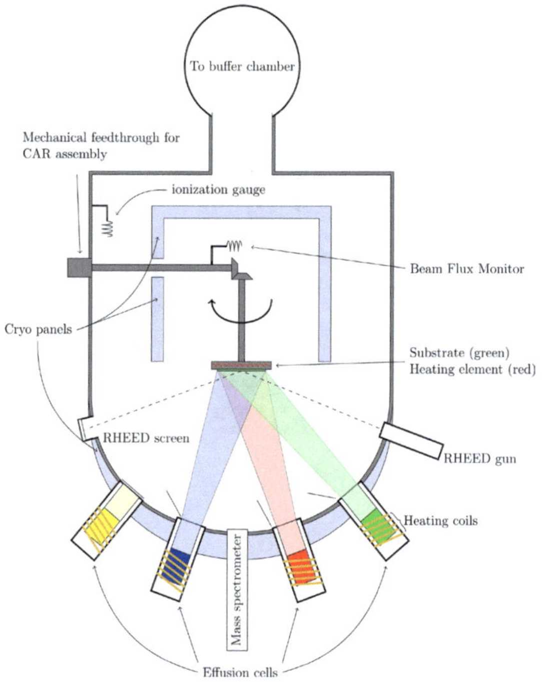

Figure 3.4. *Molecular beam epitaxy.*

Source: https://commons.wikimedia.org/wiki/File:MBE.svg, Creative Commons License.

3.6. QDS

The Stranski-Krastanov (S-K) growth mode, which is a common approach, was responsible for the self-assembling production of QDs (Placidi et al., 2012). Depending on the balance of surface energies throughout the film formation, thin films can grow epitaxially in three different ways: Frank-van der Merwe, Volmer-Weber, and Stranski-Krastanov. The material being deposited wets the substrate if the total surface energy of the film free surface and the surface energy of the substrate-film interface, is less than the surface energy, of the substrate free surface.

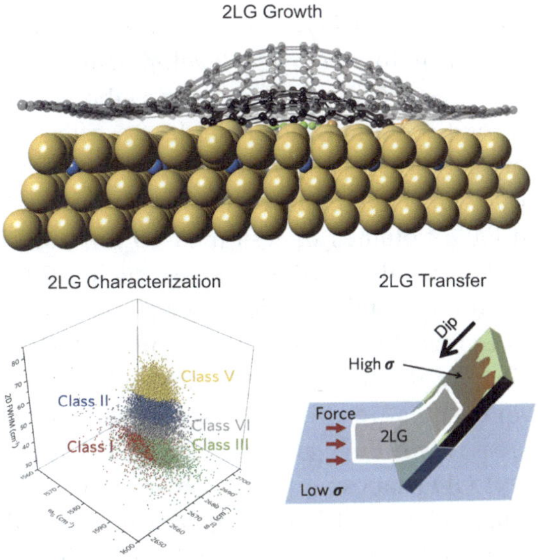

Figure 3.5. *Frank-van der Merwe growth.*

Source: Wang, H., Yao, Z., Jung, G. S., Song, Q., Hempel, M., Palacios, T., Chen, G., Buehler, M. J., Aspuru-Guzik, A., & Kong, J. (2021). Frank-van der Merwe growth in bilayer graphene. Matter, 4(10), 3339–3353. https://doi.org/10.1016/j.matt.2021.08.017, Creative Commons License.

The Frank-van der Merwe growth mode (Figure 3.5) or layer-by-layer growth, is used. It reduces surface energy by a very low lattice mismatch between the deposited material and the substrate and a strong link between the film and the substrate. The adatoms prefer to adhere directly to the substrate since they are very compatible with it. For the epitaxial synthesis of two-dimensional materials like silicene and germanene or multi-layered structures containing quantum wells, this epitaxial growth technique is utilized. Three-dimensional islands occur on the bare substrate when there is insufficient substrate bonding to prevent the film from wetting the substrate. When it comes to systems with a significant lattice mismatch, the Volmer-Weber growth mode is the usual growth mode. The Stranski-Krastanov growth mode, which is typical of heteroepitaxial systems with a nonperfect lattice match, begins with the formation of a wetting layer made up of a few monolayer-thick films (Bedoya-Pinto et al., 2021). Three-dimensional islands grow spontaneously on top of this layer. The lattice mismatch is about 7% when InAs is deposited onto GaAs. The substrate and the epilayer experience strain, and the epilayer responds to this strain by creating three-dimensional islands rather than two-dimensional surfaces. A pseudomorphic two-dimensional layer known as the wetting layer is formed by the initial few layers of InAs during development. The two-dimensional growth ceases to be energetically advantageous above a critical thickness of the wetting layer, and three-dimensional islands, often referred to as self-assembled or self-organized QDs begin to form. A barrier material, typically GaAs, InGaAs, or InAlGaAs in the InAs QDs, encloses the islands. These self-assembled dots can have volumes of less than 10 nm^3, and they are usually shaped like a pyramid. Several growth processes, such as MBE, metal-organic chemical vapor deposition, metal-organic vapor phase epitaxy, and chemical-beam epitaxy have been used to create self-assembled QDs with good crystal quality. The following are the S-K method drawbacks: The optical properties and carrier kinetics of the dots are affected by the presence of a two-dimensional wetting layer that creates two-dimensional electronic states between the QDs; the restricted range of QD shapes and sizes because of constraints on surface energies; the difficulty of controlling the QD structure and surface density separately; the limited combinations of material compositions for the substrate and dots; and the limited number of crystal orientations of the substrate.

3.7. DROPLET EPITAXY (DE)

The DE approach, (Figure 3.6) which is based on the controlled crystallization of metal nano-droplets into III-V semiconductors, is another way used to create QD structures (Sanguinetti et al., 2018). We can utilize the DE development of GaAs QDs implanted in AlGaAs as an example, but a similar process may be applied to a wide range of group III-V combinations, potentially altering the temperature ranges because of element-specific diffusivity and binding energies. Group III metals (Ga) are first deposited on the substrate surface (AlGaAs), which causes surface tension to spontaneously produce nanometer-scale droplets based on the Volmer-Weber growth mode, regardless of the lattice mismatch.

The creation of three-dimensional adatom islands is determined by the greater adatom-adatom interactions (Figure 3.7) than the interactions between adatoms and substrate. The substrate temperature can be adjusted between 150 and 500°C to manage the droplet density, which is typically in the range of 0.1 to 10^3 droplets per μm^2. The average size increases and the surface density decreases with increasing growth temperature. By carefully selecting the metal atom coverage, that is, the total amount of metal deposited per unit surface area, the droplet size, which falls between 10 and 30 nm, may be regulated. Annealing in the environment of group-V elements causes droplet crystallization, which is the next phase.

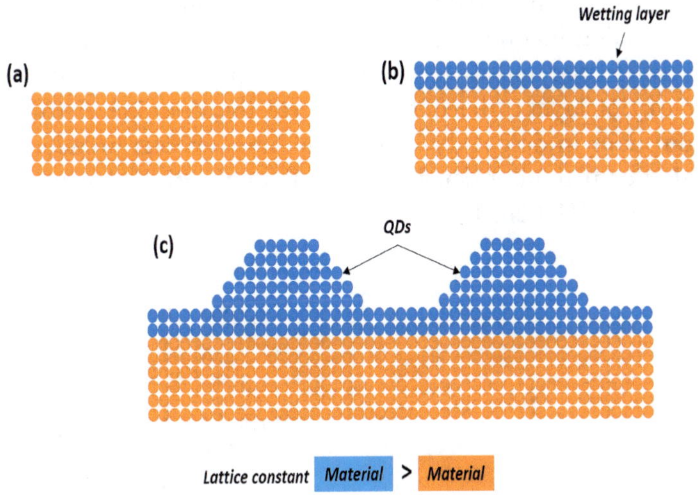

Figure 3.6. *Droplet epitaxy (DE) approach (a-c).*

Source: Gajjela, R. S. R., & Koenraad, P. M. (2021). Atomic-scale characterization of droplet epitaxy quantum dots. Nanomaterials, 11(1), 85. https://doi.org/10.3390/nano11010085, Creative Commons License.

Unlike the S-K approach, DE QDs retain their shape after capping, and by varying the substrate temperature during island formation, the size and aspect ratio may be varied widely. It can be used to fabricate both lattice-matched and lattice-mismatched QDs because strain does not drive the formation of the dots as it does with the S-K approach. The general framework of a typical QD laser includes a double heterostructure. The structure is made up of four layers of $In_{0.4}Ga_{0.6}As$ self-organized QDs as the active region, an outside cladding of $Al_{0.3}Ga_{0.7}As$, and an interior cladding of GaAs. InAs is currently the most researched QD material system for optoelectronic applications; it is frequently placed in an InGaAs quantum well, also known as DWELL (Lagraa et al., 2022).

The following are the primary benefits of QD lasers over quantum well lasers: reduced threshold current density; greater differential gain; improved temperature stability of the threshold current; reduced linewidth and linewidth enhancement factor; decreased reflection sensitivity; decreased diffusion length of nonequilibrium carriers, which results in less leakage from the active region; decreased sensitivity to defects; decreased sensitivity to nonradiative recombination centers inside or near the active region, as a result of a decreased interaction with the defects created by carrier localization in the dots. Many of these benefits have been shown in devices; however, the self-assembly methods, which are commonly employed for QD fabrication, invariably result in dot size fluctuations, causing inhomogeneous broadening, which spreads the gain over a larger spectral range and lowers the maximum peak gain. The QD density, the number of dot layers, the dot shape, and the dot size distribution affect the maximum gain in actual devices. If inhomogeneous broadening exceeds a critical value or if the dot density is less than a critical density, laser action cannot be achieved (Haken, 1966).

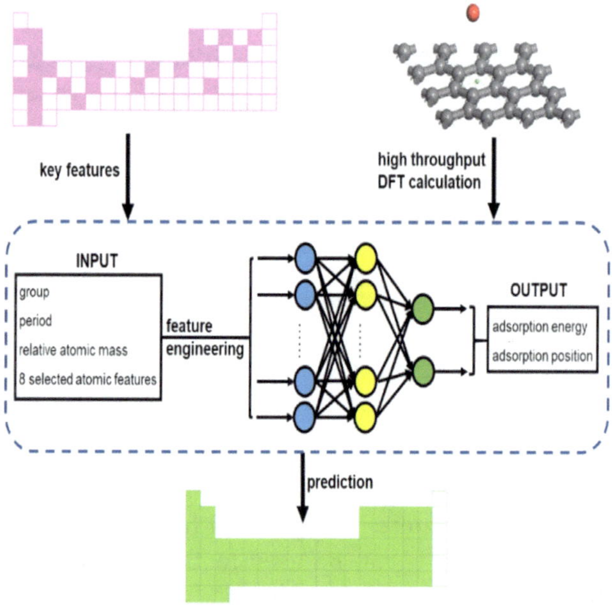

Figure 3.7. *Adatom-adatom interactions.*

Source: Qu, N., Chen, M., Liao, M., Cheng, Y., Lai, Z., Zhou, F., Zhu, J., Liu, Y., & Zhang, L. (2023). Accelerating density functional calculation of adatom adsorption on graphene via machine learning. Materials, 16(7), 2633. https://doi.org/10.3390/ma16072633, Creative Commons License.

3.8. CHARACTERISTICS

When QDs have varied shapes, such as conical or pyramid-shaped QDs, the related energy levels differ but share certain characteristics. The QD laser threshold current density can be temperature independent, meaning that its characteristic temperature, T_0, should be infinitely high. In fact, the thermal depopulation of the injected carriers across a wide range of energy states is linked to the temperature dependence of a traditional laser that is not based on QDs, which lowers the maximum gain at a given carrier injection level. Since the density of states in QDs is a δ-function, thermal depopulation of carriers disappears. As long as the electron population at higher energy levels can be taken to be insignificant, the temperature dependency totally disappears. Although actual QD lasers have better temperature stability than quantum well lasers and a very high but finite characteristic temperature, threshold current density or j_{th} is temperature dependent. For a device to have temperature-independent j_{th}, two requirements must be met: charge neutrality should be confirmed and the entire injection current should recombine in the QDs. The optical confinement area, which is the waveguide material into which carriers are introduced prior to reaching the QDs, is where QDs are always inserted in a laser (Chow and Jahnke, 2013). A component of the threshold current density is produced by carrier recombination in the waveguide (barrier) region. This element helps with the thermal depopulation of carriers from QDs, which is temperature-dependent. In the situation of equilibrium where $T > Tg$, this is the source of the temperature dependency of the threshold current density. When filling occurs in nonequilibrium ($T < Tg$), j_{th} is nearly temperature independent. The temperature dependence of the bimolecular recombination coefficient, B, of the thermal velocities of electrons and holes, and of the cross-sections of carrier capture into the QD, σn,p, is the cause of the weak dependance on T (Asryan and Suris, 2004). The occupancy probabilities are entirely temperature independent if charge neutrality is expected. At temperatures when thermal depopulation of carriers from the QDs and recombination outside the dots are inhibited, this is the cause of the temperature dependence of j_{th} and of finite values of T_0. The highly asymmetric conduction and valence band offsets in commonly used material systems, InAs/GaAs and InAs/InP, are a significant factor that restricts the temperature characteristics of QD lasers. This results in an unfavorable offset in the valence band, making the hole states much less confined than the electron states. P-type modulation doping in the QD active region can reduce the problem by counteracting the impacts of thermalization. In terms of high temperature CW operation and characteristic temperature, T_0, QD lasers generally exhibit better temperature characteristics than their quantum well equivalents, with high temperature CW operation reaching 220°C.

3.9. DENSITY

The symmetric density of states of QDs should lead to a disappearing α_H factor based on the Kramers-Kronig relations (Bimberg et al., 2001). At higher injection levels, changes in the real and imaginary components of the refractive index are dissociated due to the gain being decoupled from the carrier reservoir, resulting in a low ratio of $\Delta nr/\Delta g$. Very small values of αH, 0.1, have been measured in QD lasers, although normal values of αH in quantum well lasers fall between 3 and 5 around threshold (Fathpour et al., 2005). In addition to reducing linewidth, a low αH also lowers the laser sensitivity to undesirable optical feedback brought on by waveguide crossover, unintentional reflections from different interfaces, etc. Unwanted feedback can lead to negative consequences like mode hopping, linewidth broadening, and enhanced amplitude noise. On GaAs substrates, InAs/GaAs QD lasers have been found to be significantly less sensitive to feedback than quantum well lasers. Multimode operation or even complete coherence breakdown can be brought on by high feedback levels. The extremely low sensitivity to crystalline defects due to in-plane quantum confinement, the reduction of the diffusion length of carriers outside the QD, and the mechanical hardening of the crystal lattice in the vicinity of the dots is another significant benefit of QD lasers over quantum well lasers. Carriers that may otherwise disperse toward a dislocation to recombine are effectively trapped by the strong confinement of carriers in certain directions. The probability of charge carriers locating a dot prior to a dislocation is very high, given a QD density of up to 6×10^{10} cm^{-2} and normal dislocation densities of about 10^6 cm^{-2} in optimized buffers. Dark-line defects (DLDs), which are usually dense three-dimensional networks of dislocation loops forming around a mismatch dislocation spanning the active region, are frequently linked to semiconductor laser failure (Norman, Jung, et al., 2021). The interaction between point defects and minority carriers injected into the laser active region by electric injection and optical generation as a result of the laser light self-absorption within the laser cavity causes the growth of DLDs, which is a process of dislocation elongation. The development and rapid expansion of large DLDs are linked to the deterioration of laser performance. With a dislocation density of up to 7×10^6 cm^{-2}, recent developments in QD laser production have resulted in degradation-free functioning at room temperature (Norman, Mirin, et al., 2021). The grating structure, which facilitates the coupling between forward and backward waves in a DFB or distributed feedback laser, is created along the laser cavity length. Two Bragg reflectors are positioned at either end of the laser cavity in a Distributed Bragg reflector (DBR) laser. Two passive gratings with regularly varied refractive indices make up the Bragg reflectors. One significant benefit is that the grating is accomplished in a section of the apparatus that is physically isolated from the active area. The production processes are made simpler by this feature. However, because the Fabry-Perot cavity contains multiple longitudinal modes, a DBR laser frequency selectivity is usually inferior than a DFB laser. The structure's high reflectivity band usually contains a single mode because of the extremely low cavity length, although temperature changes may cause mode hopping. DBR lasers are therefore utilized less often than DFB lasers.

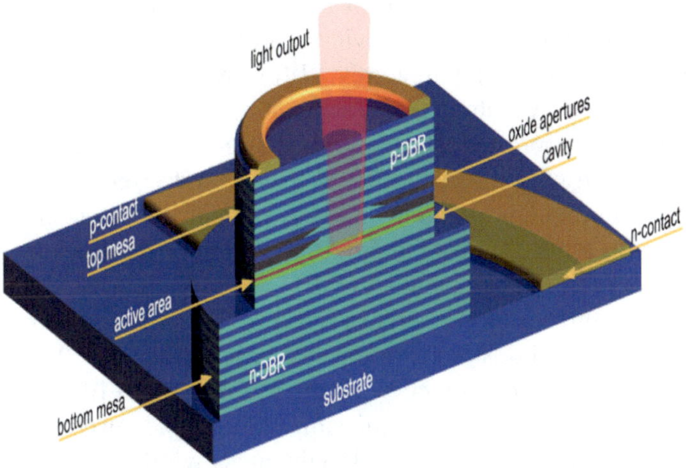

Figure 3.8. *Vertical cavity surface emitting lasers.*

Source: Tian, S.-C., Ahamed, M., & Bimberg, D. (2023). Progress in short wavelength energy-efficient high-speed vertical-cavity surface-emitting lasers for data communication. Photonics, 10(4), 410. https://doi.org/10.3390/photonics10040410, Creative Commons License.

Without specialized optical equipment, edge-emitting lasers usually oscillate in multiple modes and produce an output beam with considerable divergence. A varied laser geometry can be used for several applications, especially when the laser beam needs to be coupled into a single-mode optical fiber. The laser beam travels in a direction perpendicular to the junction in a vertical cavity. These are called as Vertical cavity surface emitting lasers (VCSELs) (Figure 3.8). These lasers are significant for having short active medium length, which results in a very minute gain. The semiconductor material epitaxial growth axis is where the laser cavity forms. To minimize internal losses, the minimal gain necessitates end mirrors with extremely high reflectivity. Bragg reflectors are utilized, which arc achieved by epitaxially depositing alternating layers with low and high index of refraction with regulated thickness, because the Fresnel reflection at the semiconductor-air interface is insufficient to ensure adequately minimal losses in the cavity. The reflectivity of the bottom and upper mirrors must be extremely high, at least on the order of 99.5%, in order to generate a suitable threshold current because the active medium is short. Because of this, DBRs must be used instead of metallic mirrors. The DBRs utilized in VCSELs are produced by alternating quarter-wavelength thick layers with low and high refractive indices (Babichev et al., 2023). Effective current confinement structures are essential for lowering the threshold current. Top or bottom emitting proton-implanted devices can be employed to restrict carrier diffusion. The semi-insulating area created following the implant confines the current, which then flows into the active area. Based on this construction, which ensures good current confinement, a number of commercial devices are made. Proton implantation may result in optical losses, and laser dependability may be impacted by implantation damage. These optical losses usually restrict the laser efficiency to a maximum of about 20%. A planar fabrication technique is made possible by the ion implanted VCSEL. Since there are no lateral photon confinement structures and the active region extent is dictated by the current confinement, these methods are examples of gain-guided VCSELs. The active region injected carriers may disperse, lowering the laser efficiency. It is easy to achieve single transverse mode emission with simple current injection. The etched mesa in which current is limited to the mesa width, is another frequently used arrangement. The etch depth is of a few micrometers, and the production process is nonplanar. Several dry etching methods, such as chemically assisted ion-beam etching and reactive ion etching, which is based on the usage of $SiCl_4$ and can yield etch depths of the order of 10 μm, can be used to create mesa structures. Dry etching has the benefit of enabling the creation of structures with extremely small diameters. To ensure dependable device operation, etching is stopped well above the active layer. Good current confinement and an optical waveguide for the released photons are both provided by the mesa structure.

3.10. VCSEL

Since the first VCSEL device was proposed in 1977 and the first operation was demonstrated in 1979, VCSELs have become widely used laser devices for a variety of applications (Koyama, 2014). Because of their symmetric mode profile and small beam divergence, they have replaced edge-emitting lasers in applications where laser-optical fiber coupling is crucial. Top-emitting VCSELs can reach coupling efficiency of up to 90% in single-mode optical fibers. Temperature changes have limited effect on the emission wavelength and threshold conditions. Emission in a single longitudinal mode is simple. Chip manufacture can be made at extremely cheap costs by using monolithic techniques to fabricate a variety of laser devices. The VCSELs enable optical on-wafer testing after growth, which is comparable to testing integrated electronic circuits. It is possible to create closely spaced two-dimensional laser arrays. VCSELs have significant uses in optical fiber communications since they can be modulated at gigahertz frequencies because of their low cavity round-trip time (Dalir et al., 2014). The quantum cascade lasers (QCLs) (Figure 3.9) are unipolar devices because they only use one type of charged carrier and the radiative recombination between two bound states of a potential well occurs within a single band (Jirauschek and Kubis, 2014). An electron can tunnel into a nearby well to repeat the emission process after undergoing a quantum jump between two levels in a specific quantum well with the emission of a photon: We refer to this process as electron recycling. As an electron travels along the laser structure, it can emit many photons.

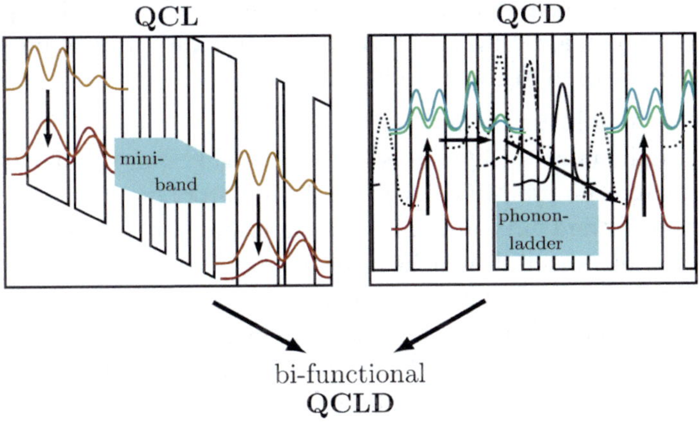

Figure 3.9. *Quantum cascade lasers.*

Source: Schwarz, B., Reininger, P., Detz, H., Zederbauer, T., Andrews, A. M., Schrenk, W., & Strasser, G. (2013). Monolithically integrated mid-infrared quantum cascade laser and detector. Sensors, 13(2), 2196-2205. https://doi.org/10.3390/s130202196, Creative Commons License.

Infrared spectroscopy, which measures the transmittance or absorbance spectrum of an infrared laser beam after it has passed through a sample, is a significant use case for QCLs. Since the majority of molecules have their fundamental vibrational modes in the mid-infrared spectral region, which lies between 3 μm and 15 μm, this region is especially important. Applications for chemical sensing include industrial process monitoring, medical diagnostics including breath analysis, and remote sensing of atmospheric pollutants and environmental gases. It is crucial to use single-mode, narrow-linewidth, adjustable mid-infrared lasers for all of these applications. This is necessary to accomplish high sensitivity and selectivity multiple-species detection. QCLs, which naturally emit in the mid-infrared spectrum, can be used for this. Several distinct cavity configurations have been used to show the single-mode and adjustable operation of QCLs. A recent example is a QCL source that can be adjusted between 6.2 and 9.1 μm. It was achieved by combining an on-chip beam combiner with an eight-laser sampling grating DFB laser array (Zhou et al., 2016). The DFB QCL ability to continually adjust its emission wavelength by adjusting the injection current or heat sink temperature, which modifies the device temperature, is one of its key features. Both situations result in a shift in the resonance wavelength of the Bragg grating of the DFB structure due to a change in the effective refractive index, which has a typical tuning rate of about 0.1 to 0.2 $cm^{-1}K^{-1}$. This is crucial since the DFB QCL offers extremely high trace gas detection sensitivity by continually tuning across the absorption peaks of different compounds. Several DFB QCLs with various grating times on a single chip can be used to cover a wide wavelength ranges.

Free-space communication is another use for QCLs. The benefit is the ability to use a wavelength in the 8–13 μm or 3.3–4.2 μm highly transmissive windows of the environment. The use of QCLs as local oscillators in heterodyne detection systems for the far and mid-infrared spectrum is another important application. Many uses in this spectral area have been made possible by advancements in terahertz QCLs. Using QCLs as a high-average power laser source for terahertz imaging is crucial. For bulk semiconductors, epitaxial growth of one semiconductor onto another is only feasible if the lattice spacings of the two semiconductors are nearly equal, with the highest relative differences being of the order of 0.1% of the lattice constant. Semiconductor heterostructures, or structures made up of semiconductors with various energy gaps, can be created in this manner. Since the lattice period of GaAs (564 pm) is nearly identical to that of AlAs (566 pm), a GaAs/AlGaAs heterostructure can be readily formed (Ringel and Grassman, 2011). Heterostructures with a specific energy gap offset can theoretically be created by combining semiconductors. By employing ternary or quaternary alloys, the range of band gap offsets can expand, providing the opportunity to customize the device operational wavelength. A lattice mismatch causes the semiconductor to experience mechanical stress, which can lead to the creation of dislocation defects at the interface and negatively impact the optical and electrical characteristics. The two semiconductor lattice constants may differ greatly as a result of the QW layer extremely thin thickness. Strain forms inside the QW layer. Strain alters the QW semiconductor characteristics, which could have significant advantages for certain applications. The optical properties of QWs differ significantly from those of bulk semiconductors due to the quantum confinement of electrons and holes in QWs and, for strained QWs, the fluctuation in effective masses. The Fermi-Dirac distribution function provides the likelihood of occupying a specific state at thermodynamic equilibrium. It is reasonable to expect that energy relaxation between the valence and conduction sub-bands occurs very quickly when electrons are stimulated in the conduction band. Double-heterostructure lasers and QW heterostructure lasers with emission in the wavelength range of 650 nm to 800 nm are constructed using GaAs/AlGaAs heterostructures; the long wavelength limit is associated with the GaAs band gap. Tension is created at the interface as the strain energy rises with an increase in the thickness of the InxGa$_{1-x}$As layer. A mismatch dislocation occurs when this tension exceeds the tension in a dislocation line. The critical layer thickness for InxGa$_{1-x}$As deposited on GaAs is greater than ~ 10 nm for x < 0.3, below which the strain caused by lattice mismatch does not result in the formation of misfit dislocations (Liu et al., 1999). This allows for the proper design of QW devices that function across a broad wavelength range.

3.11. COMPRESSION

The semiconductor cubic symmetry is destroyed by biaxial compression in directions parallel to the interface. Lattice distortion will result in a shift in the band gap of the strained layer since the energy gap of a semiconductor is correlated with its lattice spacing. The presence of strain modifies the valence and conduction bands of strained semiconductor layers, resulting in energy shifts of the band edges and a corresponding alteration in the effective masses. The majority of semiconductor lasers that are sold commercially rely on using strained semiconductors. We need to examine the impact of strain on the optical and electrical properties of III-V semiconductors because of the exceptional practical significance. Both quasi-Fermi levels are often shifted toward the band with the lighter effective mass due to the effective mass asymmetry. The transparency density rises as a result of the asymmetry. The transparency density drops by a factor of about 2.4 in the case of a symmetric QW. The effective mass in the valence band can be reduced by using strained QWs, enabling the creation of symmetric structures. It should be noted that the transparency density is achieved in symmetric bulk semiconductors, where the density of states in both the valence and conduction bands gradually increases from zero. When an asymmetric structure is present, the situation is entirely different because the band edges do not match the maximum slope of the Fermi-Dirac function. The band edge carrier density is significantly less sensitive to changes in the location of the quasi-Fermi levels, which leads to a smaller differential gain. Using a symmetric band structure with step edges is beneficial for differential gain. Using strained QWs becomes advantageous for enhancing semiconductor device performance. We can use the single-electron approximation, which is beneficial since it enables us to use a single-particle Schrödinger equation to examine a number of electrical and optical characteristics of a semiconductor. A hole is created in the valence band when an electron is driven to the conduction band; the attraction between the electron and the hole results in a correlated motion of the electron-hole pair, which functions as an exciton. Excitons are classified into Wannier-Mott excitons and Frenkel excitons. Due to its localization to a small number of unit cells, the Frenkel exciton is delocalized in k-space and necessitates a full-band treatment for analysis. This kind of exciton is crucial in materials with low dielectric constants, like organic molecules, where screening effects are minimal. The binding energy of a Frenkel exciton is typically between 0.1 and 1 eV (Vragović and Scholz, 2003). The exciton binding energy is relatively small and the exciton dimension is larger than the lattice spacing. This is because the valence electrons in semiconductors, where the dielectric constant is large, can screen the Coulomb interaction between the electron and hole in the exciton. These Wannier-Mott excitons can be described by near band edge electron and hole states, and the effective mass approximation can be applied.

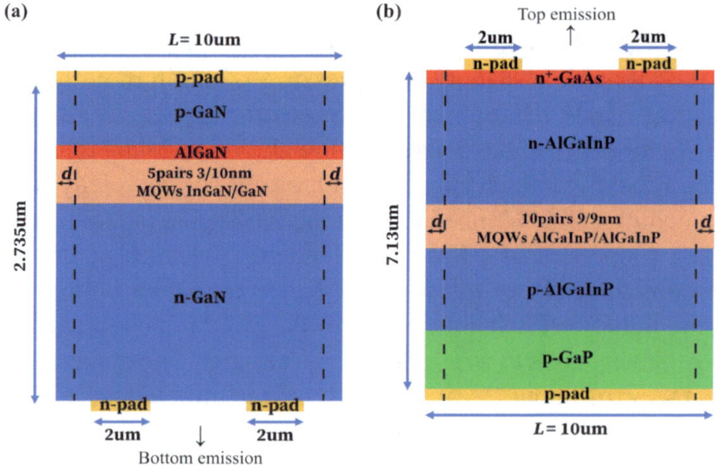

Figure 3.10. *AlGaInP double-heterostructure LED (a and b).*

Source: Ho, C.-H., Chen, S.-M., & Wu, Y.-R. (2022). Study of the factors limiting the efficiency of vertical-type nitride- and AlInGaP-based quantum-well micro-LEDs. Processes, 10(3), 489. https://doi.org/10.3390/pr10030489, Creative Commons License.

The electron and hole that make up the exciton are represented in this approximation as particles traveling in the valence and conduction bands, respectively, as per their effective masses. It is obvious that the existence of excitons alters the absorption considerably in addition to introducing distinct states in the gap. Only a small number of bound states can be spectrally resolved when the spectral broadening is taken into consideration. This is because the higher bound states combine with the ionized state absorption. A low internal quantum efficiency results from the nonradiative SRH or Shockley-Read-Hall becoming dominant term at low carrier density (Gravrand et al., 2018). This happens with a p-n homojunction LED. The holes diffuse into the n-type layer, while the electrons diffuse into the p-type layer. Since the diffusion length for minority carriers is normally between 1 μm and 10 μm, and the effective thickness of the junction is determined by Ln + Lp, the minority carriers diffuse over a wide area, lowering their density. LEDs are usually based on double heterostructures, where the active layer has a thickness of approximately 0.2–0.3 μm (Nawaz and Ahmad, 2012). This is done in order to drastically reduce the effective thickness 'd' and increase the minority carrier density. The carrier density rises at the same current density value, which raises the chance of radiative recombination and the LED efficiency significantly. A double heterostructure is composed of an active zone, where radiative recombination takes place, and two confinement layers enclosing the active region, which is distinguished by a higher energy gap. The band discontinuities and in the valence and conduction bands, respectively, must be greater than the thermal energy $k_B T$ in order to achieve an efficient carrier confinement and prevent carrier leakage into the cladding layers. Double heterostructures can be used in quantum wells, where more carrier localization is achieved in the well region, as well as in bulk semiconductors. The confinement semiconductors, which have a wider energy gap than homojunctions, do not absorb photons released in the active zone is another significant benefit of double heterostructures. The confinement areas are completely transparent to the radiation that is released since they are thin. The active zone is nearly transparent to the radiation that is released at normal carrier injection levels. The absorption edge shifts to higher energies because, carrier injection into the active layer causes the quasi-Fermi levels in the valence and conduction bands to be within the bands, ensuring that all states near the conduction band edge are populated by electrons and that transitions from the valence band to these states are impossible. We refer to this phenomenon as the Burstein-Moss shift. It is clear why LEDs do not use homojunctions (Kim et al., 2014). The thickness of the active zone is a crucial device parameter that significantly affects LED efficiency.

3.12. HETEROSTRUCTURE

An AlGaInP double-heterostructure LED, (Figure 3.10) the typical thickness of bulk active zones is one micrometer approximately (Chui et al., 1999). It is easy to understand why LED effectiveness decreases when the active region thickness increases beyond about 1 μm. When the thickness is equal to or greater than the carrier diffusion length, double heterostructures do not offer any advantages over homojunctions, and the carrier confinement caused by the double heterostructure is less effective. Significant carrier overflow occurs at moderately high levels of carrier injection if the active region thickness is reduced excessively, increasing losses. Because fewer carriers are needed to fully fill the active region, the current density for carrier overflow falls as the active zone thickness increases. When carrier overflow is enabled, increasing carrier injection does not result in an increase in radiation output; instead, it is an effective loss mechanism that causes the intensity of light emitted to reach saturation when diode current is increased. In the case of heterostructures with quantum wells, this effect is also significant. In the case of double heterostructures with low barrier heights at the interface between the active region and the confinement materials, carrier-blocking layers can be utilized to lessen the carrier leakage from the active region, which can be a significant loss mechanism. Because electrons have a generally higher diffusion constant than holes in III-V semiconductors, electron blocking layers are specifically used. A thin semiconductor layer with a high gap is positioned at the interface between the confining and active regions to form a blocking layer. The interface between two semiconductors with distinct energy gaps needs to be examined. Lower-energy states are filled by carriers in the semiconductor with a high energy gap diffusing toward the semiconductor with a smaller gap. A positively charged depletion region in the large-gap semiconductor and a negatively charged electron accumulation layer in the small-gap semiconductor form a charge separation at the interface. This charge separation results in the creation of an electric field, which participate in bending the band. Graded heterostructures, where the semiconductor composition does not change suddenly, can significantly reduce the potential energy discontinuities at the interfaces, which can raise the resistance to current flow. Not every photon produced in the active zone is released. There are several distinct processes that result in less photons at the LED output. When photons pass through semiconductor materials, they may be reabsorbed; this loss process can be reduced by employing heterostructures. When photons strike metallic electrodes, the metal absorbs them. Total internal reflection prevents a significant portion of photons from leaving the LED. It is critical to define a transmission efficiency as the ratio of photons released into free space per second to those released from the active zone per second. The ratio of photons released into free space per second to electrons injected into the LED per second is known as external quantum efficiency, and it is calculated by multiplying the transmission efficiency by the internal quantum efficiency. The majority of the light rays that are released outside the cone are absorbed by the substrate after being trapped inside the semiconductor.

CHAPTER 4

SEMICONDUCTOR LASERS – SINGLE FREQUENCY

Contents

4.1. INTRODUCTION

High coherence, including temporal and spatial coherence, is one of the primary characteristics of lasers (Salières et al., 1999). When we calculate the optical power per unit area and per unit solid angle, it denotes a high beam quality close to the diffraction limit, or high brightness. The latter denotes a very narrow linewidth, or high monochromaticity. In contrast to light from the sun and incandescent lamps, light resolved by a prism or grating is monochromatic. The lines of radiation produced by atoms or molecules changing energy levels are substantially smaller than those produced by a standard monochromator. The line becomes much narrower as a result of the laser oscillation. It has now been possible to create lasers with frequency linewidths smaller than 1 Hz. Linewidth , or frequency width is used to quantify the monochromaticity of a laser. The coherence time , which is proportional to the reciprocal of , is another characteristic that defines it. The distance that light travels over the duration of is known as the coherence length L_c, and it is written as follows:

$$\tau_c = \frac{1}{2\pi\delta v}$$

is the speed of light in a vacuum. The optical interferometer performance demonstrates the direct impact of coherence. The coherence length of the source is the largest usable optical path difference, or OPD; beyond Lc, the interference fringes grow increasingly indistinct, reducing resolution and precision. A laser mode structure affects its linewidth. Two cavity mirrors cause laser oscillation. A standing wave optical field is created, and longitudinal modes produce a spectrum when the optical channel is perpendicular to the mirrors and the phase shift between roundtrip propagations is an integer multiple of 2π. The gain spectrum and mode selectivity process typically decide that the mode amplitudes are not equal.

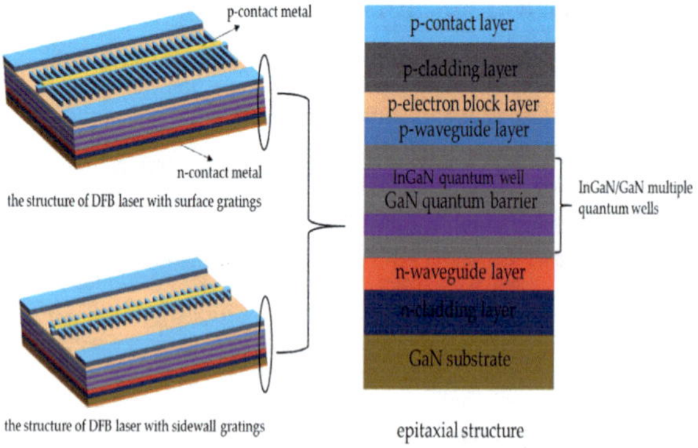

Figure 4.1. *DFB laser.*

Source: Deng, Z., Li, J., Liao, M., Xie, W., & Luo, S. (2019). InGaN/GaN distributed feedback laser diodes with surface gratings and sidewall gratings. Micromachines, 10(10), 699. https://doi.org/10.3390/mi10100699, Creative Commons License.

A mode is referred to as the main mode if its amplitude is much greater than the others; the side mode suppression ratio (SMSR) is the ratio of that mode amplitude to that of the second highest side mode. The laser is thought to operate in a single longitudinal mode if the SMSR reaches a high level that is necessary for the application, such as greater than 10 (Socorro et al., 2016). Depending on the cavity structure, the optical wave may travel in the cavity in directions with varying angles to the mirror, forming various field distributions known as transverse modes. The fundamental mode has the smallest angle, or a propagation path parallel to the cavity length. Higher-order transverse modes will have a different lasing wavelength than the fundamental mode. The associated transverse modes will increase the longitudinal mode linewidth. A narrow line signal frequency laser needs to operate in both the fundamental transverse mode and a single longitudinal mode. The laser phase noise is reflected in its linewidth. The laser beam is inherently prone to spontaneous emission. The ratio of spontaneous emission to stimulated emission will decrease as the linewidth narrows as the pump to the laser increases.

4.2. SEMICONDUCTOR

The 1960s marked the beginning of semiconductor laser development; in the 1970s, the laser threshold was significantly lowered, and it was possible to achieve CW operation at room temperature (Eichler et al., 2018). In 1980s, the design and construction of the laser cavity for the basic transverse mode was made. The semiconductor laser entered a phase of large-scale industrialization after numerous experiments with longitudinal mode control schemes. The DFB laser, (Figure 4.1) which performs in a dynamically single longitudinal mode, was chosen as the primary light source for optical fiber communications. Optical communications are coherent. Computers and optical communication enable information technology (IT) to advance. Coherent optical communication significantly increases communication capacity and speed by using the optical wave frequency, phase, and polarization state in addition to its amplitude as information carriers. High coherence, reduced noise, and improved laser source stability were needed. Cold atomic physics research which was one of the significant developments in laser technology is the cooling of atoms by laser, which raises the frequency and time standards. Another significant advancement in laser technology, the optical clock, which uses the optical frequency comb as the mechanism of frequency counting, emerged as a result of the invention of the cold atomic clock. For their contributions to the advancement of laser-based precision spectroscopy, especially the optical frequency comb, J.L. Hall and T.W. Hansch were awarded the Nobel Prize in Physics in 2005 (Hänsch and Picqué, 2012). The Michelson interferometer measurement of light velocity in ether served as one of the experimental foundations for Einstein relativity. Since then, interferometry has been widely employed in many different fields; for example, it is used to measure inertial waves. Scientists and are becoming more interested in interferometers that use semiconductor lasers as the source. One common example is the hydrophone and geophone, which are employed in a variety of fields, including navigation, fisheries, oceanography, seismic wave detection, and geoscience. Because the cavity length is shorter and the cavity mirror reflectivity is significantly lower than that of gas lasers and solid state lasers, the linewidth of an ordinary semiconductor laser is often larger. The DFB laser linewidth has been lowered to a few megahertz. The linewidth needed for high precision interferometer and spectroscopy is less than kHz, even in the order of Hz.

Reducing the linewidth of a single frequency laser is its first task. Designing and constructing an external cavity laser with a longer cavity length and a higher cavity mirror reflectivity is the common technique. Frequency noises and source drifts will reduce the accuracy of related equipment in laser cooling and technologies involving cooled atoms, precision laser spectroscopy and interferometry, and many other related applications. The laser frequency needs to be precisely adjusted and maintained at a certain material absorption line. Linewidth broadening methods are used in frequency stability. In 1981, A.L. Schawlow was awarded the Nobel Prize in Physics for developing saturated absorption spectroscopy. The ultra-stable Fabry-Perot cavity, or PDH technique, developed by R.V. Pound, R.W.P. Drever, and J.L. Hall, allowed for frequency stabilization down to Hz and even lower (Li et al., 2024).

4.3. LASER FREQUENCY

In order to be used efficiently, the laser frequency must also be swept throughout a range quickly, linearly, or in a specific waveform, while maintaining a sufficiently thin linewidth. The ability to convert the laser frequency to any other frequency is also necessary for some applications. Due to the direct interaction between photons and electrons injected by current, as well as the semiconductor vulnerability to external disturbances, the noise level of semiconductor lasers is significantly higher than that of gas and solid state lasers.

One of the advantages of the single frequency laser is noise reduction, which includes frequency stability and frequency noise. Various designs were used to stabilize the laser frequency for this reason, and new technologies are being created. The noise must be measured, particularly the noise spectrum. The optical phase should be locked between lasers within a certain error level; this leads to the development of optical phase locked loop (OPLL) technology

(Figure 4.2) (Balakier et al., 2018). One of the sources of noise in many laser applications, including interferometers, is polarization variation. One of a single frequency laser capabilities is a stable and regulated polarization state.

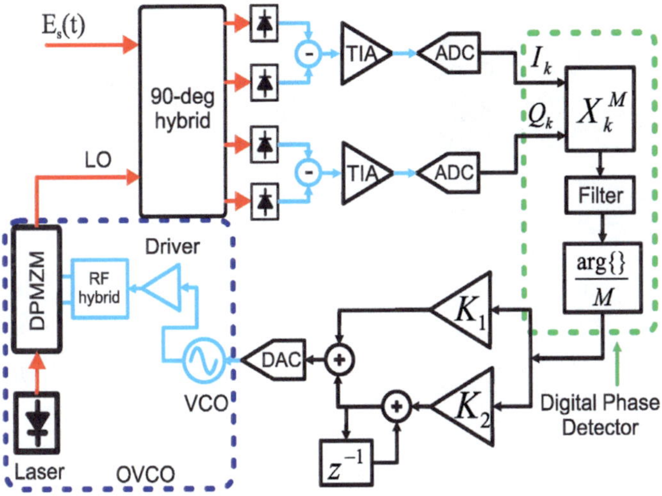

Figure 4.2. *Optical phase locked loop (OPLL) technology.*

Source: Panasiewicz, J., Arab, N., Destic, F., Pacheco, G. M., & Rissons, A. (2023). An all-digital optical phase-locked loop suitable for satellite downlinks. Photonics, 10(12), 1312. https://doi.org/10.3390/photonics10121312, Creative Commons License.

Simultaneous beam quality control and enhancement are necessary. Research on the angular momentum of photons, particularly the orbital angular momentum (OAM) of light, has gathered interest in recent years. The development of the semiconductor laser is ongoing. Physical mechanisms are being discovered, new experimental phenomena are being perceived one after the other, its performances are being quickly improved, and its applications are growing extensively. Transitions between the conduction and valence bands, or the emission via inter-band recombination of carriers, are the cause of light emission in semiconductors. Nonequilibrium carriers, such as electrons in the conduction band and holes in the valence band, are the active particles in semiconductors. The Fermi-Dirac distribution is followed by the electron population. The emission and absorption rates should be computed by integrating the energy with the state density since the electrons and holes are dispersed throughout the sub-energy levels of the bands. The material structure and the physical conditions, including temperature, affect the state density function.

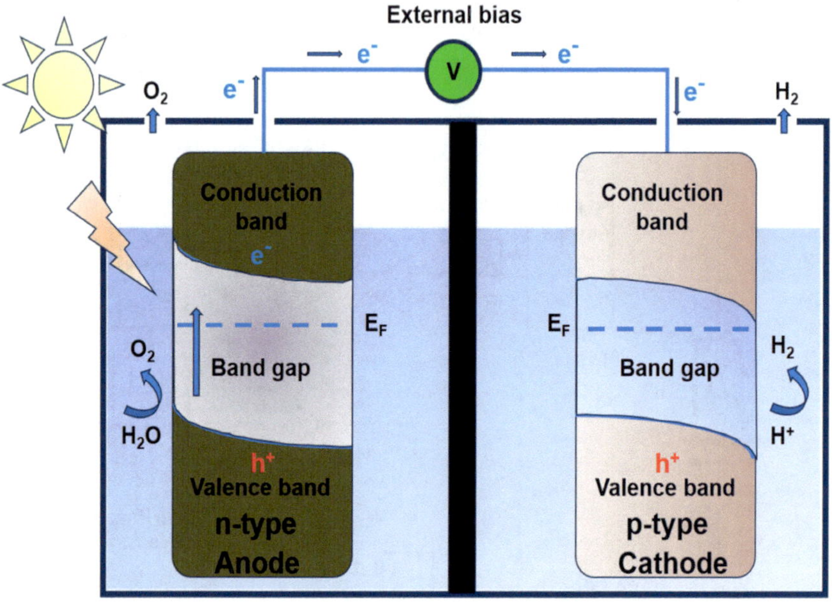

Figure 4.3. *Semiconductor band structures.*

Source: Pech-Rodríguez, W. J., Şahin, N. E., Suarez-Velázquez, G. G., & Meléndez-González, P. C. (2025). Semiconductor-based photoelectrocatalysts in water splitting: From the basics to mechanistic insights—A brief review. Materials, 18(9), 1952. https://doi.org/10.3390/ma18091952, Creative Commons License.

Semiconductor band structures (Figure 4.3) are classified into two groups: direct bandgap and indirect bandgap. When the momentum conservation law is satisfied, the transition in direct bandgap materials takes place close to the wavevector space. The indirect bandgap valley of conduction band is located far from the zero point. A phonon, which is the quantized particle of crystal lattice vibrations and has a more momentum and a lower energy than photons, must be included in the transition in indirect bandgap materials. The majority of III-V compounds meet the requirements for lasing and are classified as direct bandgap materials. Research studies shows that the bandgap, or difference in quasi-Fermi levels, is equivalent to the photon energy of a semiconductor laser. Several semiconductors and their solid solutions with various compounds can be used to create semiconductor lasers in a wide spectral band, from ultraviolet to middle infrared. In addition to III-V compounds, several compounds in the IV-VI group, such PbS, have direct bandgaps and are utilized in lead-salt semiconductor lasers that operate in the middle infrared bands (Zogg and Ishida, 2001).

4.4. P-N JUNCTION

In semiconductors, lasing requires injecting carriers with a density high enough to allow the quasi-Fermi levels to enter the valence band and the conduction band, respectively. The optical pump is one technique; the other is electron beam injection (vacuum-induced). These approaches are not practical and have very low efficiency. Another method for injecting high concentration carriers is the semiconductor P-N junction. P-type and N-type zones with matching lattice constants make up a P-N junction. Carriers can travel between the two areas or regions by diffusion under a carrier concentration gradient as well as drifting under an electric field. When there is no voltage applied, electrons and holes spread across the P-N junction interface in opposite directions. Two carrier-depleted layers with discharged positive donors and residual negative acceptors form on either side of the junction. The energy bands of the two locations alter as a result of the inherent Coolum field balancing the carrier diffusion. The P-region energy level decreases when a positive bias is applied; the positive injected current is produced by the injection of electrons from the N-region into the P-region and holes in the opposite direction. The nonequilibrium carriers are concentrated in two regions close to the junction; photons are released as a result of the nonequilibrium carriers recombining. The Fermi levels close to the junction are now divided into two quasi-Fermi levels that represent electrons and holes, respectively. The injected carriers accumulate higher as the applied voltage moves closer to the bandgap, causing the quasi-Fermi levels to move toward the top of the valence band and the bottom of the conduction band, respectively. This is known as stimulated amplification of emission. The majority of widely used near-infrared lasers operate at a voltage of about 1 V. Since the series resistance occurs independently of the junction voltage and includes the Schottkey barriers on the metal-semiconductor contact between the laser material and electrodes as well as the resistance of volume materials, the actual voltage measured on the electrodes will be slightly greater than the bandgap. One benefit over other lasers is their low operating voltage. The laser should be constructed from highly doped low resistivity materials, which offer the advantages of both a larger injected carrier concentration and a low series resistance. The quasi-Fermi levels are simpler to move in order to meet the lasing condition since the equilibrium Fermi levels are near the top of the valence band and the bottom of the conduction band, respectively. In addition to semiconductors with identical lattice constants, materials with slightly differing lattice constants can also be used to create the quantum well. In the latter, the junction is conserved as a mono-crystal under specific conditions, while the quantum well undergoes asymmetrical deformation in directions perpendicular and parallel to the layer. The material is referred to as strained layer quantum wells, which include that which is stretched in the plane and compressed in the plane of crystal development. The lattice matching restriction was overcome by the development of strained layer quantum well, which significantly extended the semiconductor laser spectrum range.

4.5. ENERGY BAND

The energy band structure changes as a result of the strains, and new characteristics are developed. Energy band engineering is the process of designing and creating strained layer quantum wells. A novel type of semiconductor laser called a quantum cascaded laser (QCL), which uses transitions between sublevels of the quantum well rather than transitions between the valence band and the conduction band, is developed. The wavelength range is further extended, particularly in the infrared band. The standard LD based on inter-band transition may be hard to function at room temperature due to the high concentration of thermally excited equilibrium carriers, which prevents the population inversion from occurring in materials with such a low bandgap. Most of the semiconductor lasers and LEDs that are currently on the market are commercially accessible and range from the ultraviolet to the middle infrared spectrum. p and n-type claddings are the upper and lower layers of the active zone, which is a thin core layer of an optical waveguide with a typical thickness of less than 1 μm. In order to lower the pump current and preserve the essential transverse mode, the active zone is restricted by burying materials with a lower index but with high resistivity, creating a stripe that is only 2 to 5 μm wide. The two cleaved facets of the semiconductor crystal make up the cavity. III-V compounds with cubic crystal structures make up the majority of semiconductors used in lasers (Cheng, 2020b). The cavity length falls between 0.3 and 1.0 mm. The surfaces are where the electrodes are constructed. With a thickness of roughly 0.1 mm and a width of less than a millimeter, the LD chip demonstrates its benefits of being compact and light. Liquid phase epitaxy (LPE), metal organic vapor phase epitaxy (MOVPE), or molecular beam epitaxy (MBE) are used to build the several layers on the substrate. With the exception of the cavity mirror created by cleaving and the output via windows or fiber coupling, the material growth, device fabrication, and packaging processes are similar to those of standard microelectronic devices. The semiconductor laser offers the benefit of low cost and is appropriate for mass production. The heterostructure functions as a planar optical waveguide; the P-N junction is included in the core, which is a smaller bandgap but with higher index layer, while the cladding is made of materials with a huge bandgap and lower index. The transverse modes in dimension perpendicular to the junction are determined by the waveguide. The majority of semiconductors operate in TE modes, which involve an electric field in the P-N junction plane. The TE modes have a lower threshold because their facet reflectance is slightly higher than that of TM modes. As the output power rises, so does the degree of polarization. A TM mode laser can be achieved by creating a stained layer quantum well with a gain coefficient of TM mode that is higher than that of TE mode using specific designs. Only the fundamental mode with the highest β operates because the core thickness is so small (1 μm), while all higher order transverse modes are suppressed.

4.6. MODES

Multiple transverse modes in the junction plane dimension operation is simple, which is undesirable for the majority of applications. Since the cavity of gas and solid-state lasers is made up of optical parts, the transverse mode can be created. The lateral boundary of the LD cavity, which is composed of cleaved facets, must be created by semiconductor processing during the material growth and/or chip fabrication stages. A larger bandgap semiconductor buries a stripe mesa that forms in the burred heterostructure. The widely employed structure is most likely the ridge waveguide, in which the thickness of the top cladding outside the stripe is reduced during processing, while the thickness above the stripe stays the same, creating an effective index difference between the two areas. The stripe width and index difference should be small, such as 2–5 μm and less, to guarantee that the laser operated at the fundamental lateral mode. The beams emitted from the laser cavity will have varied properties due to the variable wavefront inside the waveguide. A Gaussian beam with distinct divergences in directions parallel and perpendicular to the junction plane that correspond to the modes with a lower and a higher effective index roughly characterizes the beam. The locations of the beam waist for index guiding and gain guiding differ in the slow axis. The beam waist is only at the facet, and the wave is directed by a pure index distribution in the fast axis. Where the astigmatism is very small in index-guided lasers, it is present in gain guide lasers. The output beam divergence angles of LD are rather large, usually between $10 \times 40°$, due to the fundamental mode diameters (small) (Murray et al., 2006). This is larger than that of gas lasers and solid-state lasers, which are on the order of milliradian. In LD applications, an optical collimation system is typically required; the optics design must take into account the beam properties, particularly its astigmatism. Its ability to couple with the optical fiber is an added benefit. There are several applications that require the highest feasible laser power, including medical applications and solid-state laser pumps. A high-power laser is made up of several broad area stripes to create an array, or even a stack of LD array bars. Optical phase locking results from optical coupling between stripes, which happens when the distance between them is reduced to a particular extent. The phased array is the term given to such a type of array. The super-mode theory must to be used to examine its mode behavior. The zero order super-mode, whose output beam is a single narrow lobe, is the mode with zero phase difference between neighboring stripes. The highest order super-mode with an output beam of two narrow lobes is the mode with a π phase difference between consecutive stripes. The threshold is typically lowest for the highest super-mode. Phase locking significantly reduces the divergence angle. For high power lasers, however, thermal performance is crucial; a large distance between stripes helps to lower temperature rise at high pump current. LD stacks, which are appropriate for pulsed operations, are also created to achieve higher power.

4.7. GRAIN COEFFICIENT

One obvious distinction between semiconductor lasers and gas and solid-state lasers is their high gain coefficient. The reason for this is that the doping concentration in semiconductors is as high as 10^{18} cm^{-3}, which is higher than the densities of active particles in solids and gases (Pereira and Almeida, 2015). Each doped atom can contribute one carrier, which raises the gain coefficient. High reflectance cavity mirrors are not required because of the high gain coefficient. Due to this, a simple cleaved facet can be used as a mirror. Fresnel calculations indicate that the reflectance of optical intensity is roughly 0.32, which is significantly less than the reflectance of other lasers, which is frequently close to 1 (Hipp et al., 2019). A short cavity length is sufficient to provide a high single trip gain because of the high gain coefficient. Ordinary LD has a cavity length of only 0.3 to 0.5 mm, whereas high power LD has a cavity length of approximately 1 mm. Large longitudinal mode spacing is the result of the low cavity length. Compared to gas lasers and solid-state lasers, the gain spectrum is wider. The reason for this is that energy bands are where the change takes place. While SLD and LED can be constructed from the same materials as LD and share structural similarities, SLD uses a very low reflecting facet and LED lacks a cavity. In order to prevent lasing oscillation, LEDs typically use an anti-reflective coating or an active stripe angled toward the cleaved facet. Because of the weak feedback from the mirror, SLD spontaneous emission is amplified, resulting in an output power that is larger than LED but still below the threshold. In certain situations, the suppression of coherent noise is aided by the significantly wider spectral linewidth of LD and SLD compared to LD. If anti-reflective coatings are applied to both of the LD active region facets to stop round-trip oscillations, the region can be used as an amplifier. This type of device is known as a semiconductor optical amplifier (SOA), and it is frequently utilized in ring-cavity lasers, optical communications, and other applications (Komljenovic et al., 2017). The gain spectrum determines the mode spacing, and the mode wavelength changes as pump levels change. We refer to this phenomena as frequency pulling. Due to their larger gain coefficient and higher dispersion, semiconductor lasers have a stronger effect than gas and solid-state lasers. A medium's linewidth broadening process affects its gain spectrum. Broadening of the linewidth is said to as homogeneous if all active atoms contribute equally; on the other hand, it is referred to as inhomogeneous if atoms contribute differently. The latter is commonly demonstrated by the Doppler broadening caused by varying atom movement velocities in gas. A homogenous broadening with a Lorentzian line shape results from the intrinsic decay and spontaneous emission of atoms in their excited state, which happens with a finite probability. The broadening in semiconductors is thought to be essentially homogeneous with a gain spectrum. The semiconductor laser operates with many longitudinal modes due to the large linewidth of the gain coefficient, particularly for the F-P cavity LD without mode selection values. The gain spectrum determines the mode intensities. The side mode suppression ratio (SMSR) will rise as the pump current increases and the main mode difference falls towards zero. It is evident that the number of lasing modes is influenced by both the working point and the mode separation, or cavity length. For the homogeneous broadening medium, the side mode intensities drop as the main mode intensity increases; this phenomenon is known as mode competition. When the pump current increases, the envelope linewidth of LD decreases. The spontaneous emission factor also affects the envelope linewidth.

4.8. EMISSIONS

A smaller γ_{sp} indicates a faster movement from many modes to a dominating main mode and a sharper transition from spontaneous emission to lasing. Experiments have shown that the index-guided laser often has fewer modes of operation than the gain-guided laser. Lasers having a single longitudinal mode are necessary for high-speed optical fiber communications, even if their linewidth might not be extremely small. There are many unique semiconductor laser structures that have been created. The most commonly utilized lasers are the vertical cavity surface emitting laser (VCSEL), the distributed feedback laser (DFB-LD), and the distributed Bragg reflector laser (DBR-LD) (Fang et al., 2017). Many significant and recently developing fields require single longitudinal mode lasers with sufficiently narrow linewidth. Other technologies are being explored for frequency stabilization and linewidth reduction. The modulation will alter the longitudinal mode behavior of the laser. Under DC current pumping, an LD with a good index guiding structure may function in a high side mode suppression ratio (SMSR) condition; but, under high frequency modulation, it will significantly degrade. The multi-mode rate equations and the gain spectrum provide an understanding of this phenomenon. The gain switch is a technique for creating short optical pulses that relies on semiconductor laser ability to modulate directly. Given a greater resonance frequency, it is anticipated that a higher injection current will be able to shorten the pulse width of the gain-switched pulses. Although the pulse width is typically not as narrow as with mode-locking, the gain switch approach has the advantages of flexibility, and low cost when compared to other short pulse operating methods, such as mode-locking. In certain situations, the semiconductor laser output may oscillate even while it operates with continuous waves.

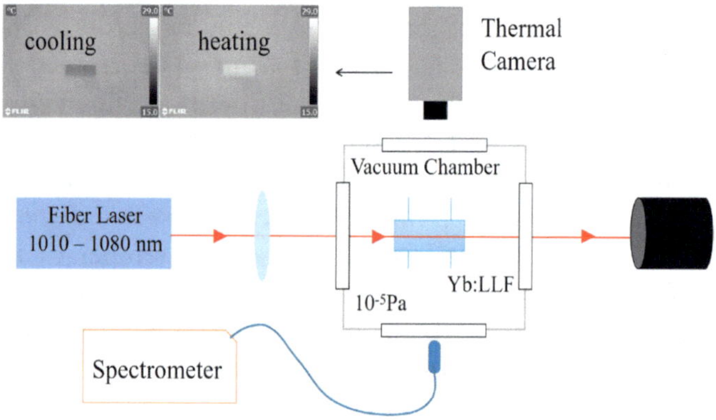

Figure 4.4. *Laser cooling.*

Source: Duan, X., Zhong, B., Lei, Y., Wang, C., Xu, J., Zhang, Z., Ding, J., & Yin, J. (2022). Accurate characterization of the properties of the rare-earth-doped crystal for laser cooling. Applied Sciences, 12(9), 4447. https://doi.org/10.3390/app12094447, Creative Commons License.

Self-sustained pulsation is the term used to describe this phenomena. It is one of the instability phenomena of lasers. The modulation frequency determines the frequency modulation response. The primary process in the low frequency region is the thermal effect of the injected current. Temperature causes the cavity length to expand, and the thermo-optic effect also causes the index to rise. Temperature modulation occurs with a delay, and the modulation amplitude diminishes as the modulation frequency rises. The primary effect in the high frequency band is the change in index with carrier concentration; the temperature of the active zone may not follow the modulation of injected current. It is observed that with current modulation, both frequency and power modulation will take place simultaneously. Power modulation, also known as residue amplitude modulation (RAM), should be avoided in many single-frequency laser applications where a pure frequency modulation is frequently required (Tull et al., 1997). Due to the inherent temperature sensitivity of semiconductor materials, LD characteristics exhibit significant temperature dependences. In terms of stability, it is a drawback. However, when viewed from a different perspective, it offers the ability to modify and regulate the properties through the temperature effect.

4.9. LASER COOLING

One of the interesting developments in laser technology is the cooling of atoms using lasers (Figure 4.4). This achievement serves as the foundation for the development of the cold atomic clock and optical clock. One important tool is the single-frequency semiconductor laser. Although it may satisfy the needs of everyday living for average people, timing with an error of less than one second per day is far beyond what modern technologies require. Even if the accuracy of a quartz clock has increased to 10^{-4} seconds per day over the past century due to advancements in electronics, higher precision is still needed for advanced scientific studies and sophisticated technology. Researchers tried to create a new frequency and time standard based on the atom absorption lines after noticing how stable the atom absorption spectra were. Atom's time, began with the development of a cesium (^{133}Cs) atomic clock (Figurc 4.5).

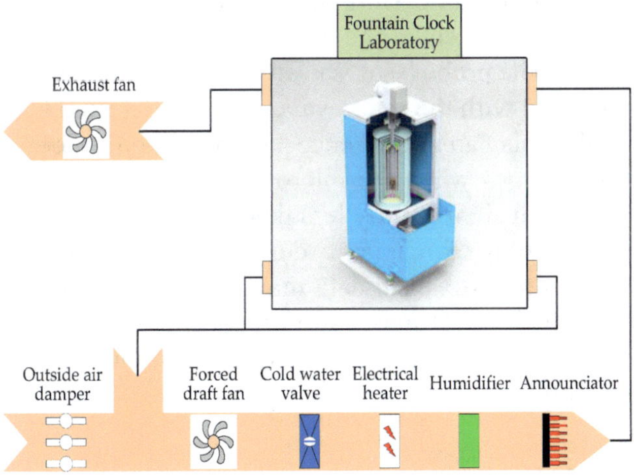

Figure 4.5. *Cesium (^{133}Cs) atomic clock.*

Source: Yang, F., Wang, X., Ruan, J., Shi, J., Fan, S., Bai, Y., Guan, Y., Hao, Q., Zhang, H., Liu, D., & Zhang, S. (2022). Experimental evaluation of the blackbody radiation shift in the cesium atomic fountain clock. Applied Sciences, 12(1), 510. https://doi.org/10.3390/app12010510, Creative Commons License.

One second was defined as the time of 9,192,631,770 cycles of electromagnetic radiation, which corresponded to the transition between the two fine levels of cesium during the 13th International Conference of Metrology in 1967 (Arora and Sen Gupta, 2023). This definition was based on the successful establishment of the atomic frequency standard. The astronomical time unit was then measured using the atom frequency standard, which supersedes the previous standard based on the sun's movement. The fundamental building block of an atomic clock is a frequency transfer chain, in which the crystal oscillator frequency is locked to the resonant frequency of a resonator that holds the vapor (atom). The atoms that are frequently utilized are hydrogen (H-maser), cesium (Cs), and rubidium (Rb). To regulate the frequency of the crystal oscillator, the beat signal is utilized as a feedback error signal. The error signal is processed using PLL, and the oscillation frequency matching with the transition frequency is cascaded using a frequency synthesizer. The sustaining time of the reaction between the microwave and the atoms increases the atomic clock precision. A long cavity is required because of the extremely high velocity of atom movement at ambient temperature. A long cavity has certain disadvantages, such as mechanical stability and microwave field uniformity. N.F. Ramsey created a unique cavity. It is a cavity made up of two pieces that are symmetrically and independently arranged, and their microwave phases are locked together. The atomic clock was successfully built and its performance was significantly enhanced by the Ramsey cavity. Ramsey received the 1989 Nobel Prize in Physics (Banerjee and Matsakis, 2023). The probe laser is used to generate the photoluminescence of the state that reacted with the microwave and was detected by a photodiode. The PD signal will display an interference pattern of atom-microwave reactions known as Ramsey fringes when the microwave frequency is adjusted. These fringes have a linewidth significantly less than that of a typical resonant cavity. By employing the action of magnetic deflection to identify the amount of reacting atoms, the interferometric data may also be measured. Many atomic clocks, like the Rb-clock and H-maser, are widely employed in many significant fields including global navigation systems and telecommunication networks for node synchronization.

4.10. FREQUENCY

The frequency width of an atom absorption line and the time that atoms and microwaves react are the two key factors that affect the accuracy of an atomic clock. The atom's fundamental properties govern the former, but thermal movement (the Doppler Effect) and the influence of an external electromagnetic field (the Zeeman and Stark Effects) will broaden it. The essential technique for improving the accuracy of the atomic clock is to cool the atoms. Utilizing cooled atoms has the added benefit of extending the microwave response time with the atoms, which lowers atomic clock noise. It was anticipated that the exchange of momentum between atoms and photons would be used to cool the atoms following the development of lasers. The cold atomic clock and the physics of cooled atoms are popular R&D topics in cutting-edge technology. In the 1990s, the cooled atom fountain atomic clock was successfully released. Steven Chu, Claude Cohen-Tannoudji, and William D. Phillips were awarded the 1997 Nobel Prize in Physics for their invention of techniques for using laser light to cool and trap atoms (Cohen-Tannoudji, 1998). In 2010, it was stated that the Cs fountain clock precision was $4–5 \times 10^{-16}$. Utilizing the cold atomic clock in space is a new goal as space technology advances, particularly with the global positioning system. It is anticipated that the frequency stability of the cold atomic clock will be improved by the gravity-free environment. In daily life, industry, and scientific activities, cooling technology has been crucial. Based on thermodynamics, the traditional approach has pushed the low temperature record well below the helium condensation point. Prior to laser cooling, there is a limit on how close one may go to the absolute zero point, which is –273.15°C (Cohen-Tannoudji and Phillips, 1990). It is commonly known that photons have a momentum of mc = hv/c in addition to an energy of hv = mc^2. When photons strike a body, they apply a pressure known as the light pressure. The light frequency and the absorption line must coincide. It is necessary to employ a laser beam with adequate monochromaticity. The excited state atom will spontaneously emit in randomly distributed directions to return to the ground state. The atoms will cool collectively if the process is accumulated. The cooling photon should have a negative detuning associated with the absorption line peak. As the atom cools, this kind of detuning is diminished. A fundamental requirement for laser cooling is a stabilized laser with adjustable tunability. Yet, the need for constant cooling cannot be satisfied by the laser frequency swept in serrated waves. Zeeman Effect or Stark Effect is used to solve the issue. Atoms' absorption frequency will alter in proportion to cooling in a spatially variable electric or magnetic field pointing in the direction of the laser beam, and it will consistently match the frequency stabilized laser as it cools.

Spontaneous emission will cause the excited state atom to revert to the background state. However, because of the electron spin, the background state is typically divided into two levels with a small gap. A small percentage of atoms that leave one of the two levels will go back to the other, increasing the total number of atoms at that level. For continuous cooling, another laser beam with a frequency matching the latter level is required. We refer to the latter procedure as re-pumping. Atoms that have cooled are trapped. Since it is evident that cooling the atoms in three dimensions is required, six laser beams that propagate in opposite directions in x, y, and z are required. Three mirrors are used to obtain the opposite beams. The molasses, which contains more cooled atoms for study, is created when the cooled atoms are concentrated around the center point. Gravity will cause the atoms in molasses to diffuse and descend. Some trapping technologies are created to maintain the cooled atoms in a limited volume. The magneto-optical trap (MOT) (Figure 4.6) is one of the most popular techniques among them. Based on the Zeeman Effect, MOT confines the atoms using a magnetic field with a gradient that has been created. Bipolar force trapping is another technique. Because of the accompanying decrease in Doppler frequency shift, the cooling process mentioned is known as Doppler cooling (Pan et al., 2021). The absorbed photon energy will convert to thermal energy when the velocity is reduced to a specific level. When the atoms move from the excited state to the ground state by spontaneous emissions, they will acquire momentum in randomly dispersed directions. There is a limit to the lowest cooled temperature and a heating effect becomes dominant at a particular low temperature.

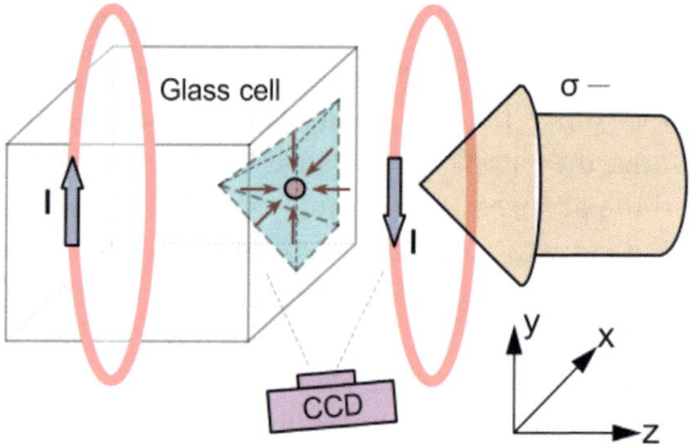

Figure 4.6. *Magneto-optical trap.*

Source: Xu, R., Li, A., Li, D., & Yan, J. (2023). Magneto-optical traps for cold atomic gravimetry: research status and development trends. Applied Sciences, 13(10), 6076. https://doi.org/10.3390/app13106076, Creative Commons License.

The limiting temperatures for cesium and rubidium are 124 and 145 μK, respectively. Sub-Doppler cooling techniques, such as velocity selective coherent population trapping (CPT), (Figure 4.7) magnetic induction, and polarization gradient, are developed to overcome this. When used to construct an atomic clock, the cooled atom will extend the reaction time with the microwave in the cavity. With the exception of vertically traveling atoms, however, the deflection of the atom beam in the gravitational field becomes more severe the slower the atoms move. It indicates that the cavity should be set up vertically so that the atomic beam moves like a fountain of water. The atoms in the atomic fountain will undergo two round-trip reactions with the microwave field; this cavity can be thought of as a folded Ramsey cavity. Six laser beams, including pump and re-pump lasers, combine to make a molasses that is trapped in the MOT that the magnetic field has created. By turning off one of the lasers, a specific percentage of the atoms in the MOT will be thrown upward and then returned. The detecting beam beneath the MOT measures the atomic state changes. A small atomic clock that makes use of diffuse laser cooling is the integrating sphere cold atom clock. The measured frequency width was only 20 Hz, which is significantly smaller than the microwave cavity width of roughly 1000 Hz.

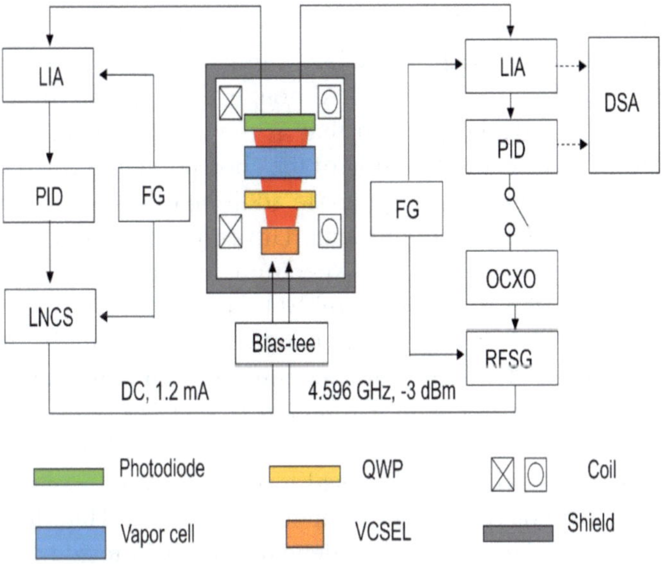

Figure 4.7. *Coherent population trapping.*

Source: Hong, H. G., Park, S. E., Lee, S. B., Heo, M.-S., Park, J., Kim, T. H., Kim, H. Y., & Kwon, T. Y. (2021). Chip-scale ultra-low field atomic magnetometer based on coherent population trapping. Sensors, 21(4), 1517. https://doi.org/10.3390/s21041517, Creative Commons License.

4.11. SEMICONDUCTOR LASER

The semiconductor laser is thought to be the option because research and development in this field does not require a high laser power demand. One of the essential components is a high coherence semiconductor laser with rapid and accurate tunability and frequency stabilization on absorption lines. It is anticipated that this will significantly improve the global positioning system accuracy. The optical clock, which uses atoms with extremely stable energy levels, like strontium (Sr), trapped in the optical lattice and combined with a femtosecond optical frequency comb for precise frequency measurement, is another project that researchers are investigating. With a precision of one second per 300 million years, the optical clock is expected to deliver the most precise frequency standards (Diddams et al., 2004). Under specific pump settings, EIT typically appears as a limited width transparency window inside a material absorption line. Its mechanism consists of a collection of energy levels, essentially in three configurations: -type, V-type, and ladder type. When the material is pumped by the coupling light, the destructive interference of the transitions between related energy levels prevents the absorption; the probe may then quantify the effect of EIT. Although the population at the ground state is kept significantly larger than that at the excited state, a transparency window will form at the central portion of the absorption line when the coupling light is on. The technique may incorporate additional levels to create a tripod-like arrangement or other configurations. Ultra-fine structural sub-levels will likewise exhibit many peaks in the EIT spectrum. According to Rabi oscillations in quantum physics, the mechanism of EIT is a destructive interference of wave functions associated with the two transitions (Bayrakli, 2021). Since no photon is absorbed or released, this condition of an atom is also known as the dark state. Coherent population trapping (CPT) is the process by which a group of atoms interact with the coupling optical field to create a trapped state. The velocity selective CPT is used to overcome the Doppler limitation in laser cooling by converting the kinetic energy into Hamiltonian. Based on Kramers-Kronig relations, it is evident that the dip inside the absorption line is quite small, producing a high dispersion of refractive index. The area has a significantly reduced light velocity. The intra-cavity EIT, which uses a slow-light effect in lasers, would efficiently shrink the laser line. In addition to slowing down atoms, the mechanical force of the laser field can capture dielectric particles as small as nanometers and micrometers in a concentrated optical beam. The optical beam can then move the particle. A unique technique known as optical tweezers is created based on this effect. The momentum of the field describes the light pressure in accordance with Maxwell electromagnetic field theory. A particle may become stuck at the laser beam focused point under some circumstances due to the centripetal force produced by the strong electric field gradient at the beam waist. Higher coherence, a regulated beam shape, and unique polarization states are necessary for the laser beam used in optical tweezers. The microparticles can also rotate due to adjustable orbital angular momentum of light.

4.12. OPTICAL COMMUNICATIONS

The development of optical communication was stimulated by the introduction of the laser. Fiber communications have advanced significantly and are now one of the pillars of the information age, particularly since Sir Charles Kuen Kao first proposed and elaborated on the use of optical fiber in communications in the 1960s. Kao received the 2009 Noble Prize in Physics for the contribution. The intensity modulation and direct detection (IM-DD) system is currently used mostly in fiber communications (de Arruda Mello and Barbosa, 2021). The data is transmitted by optical fibers, carried by optical pulses encoded with pulse code modulation (PCM) standards, and detected by a photodetector at the terminal.

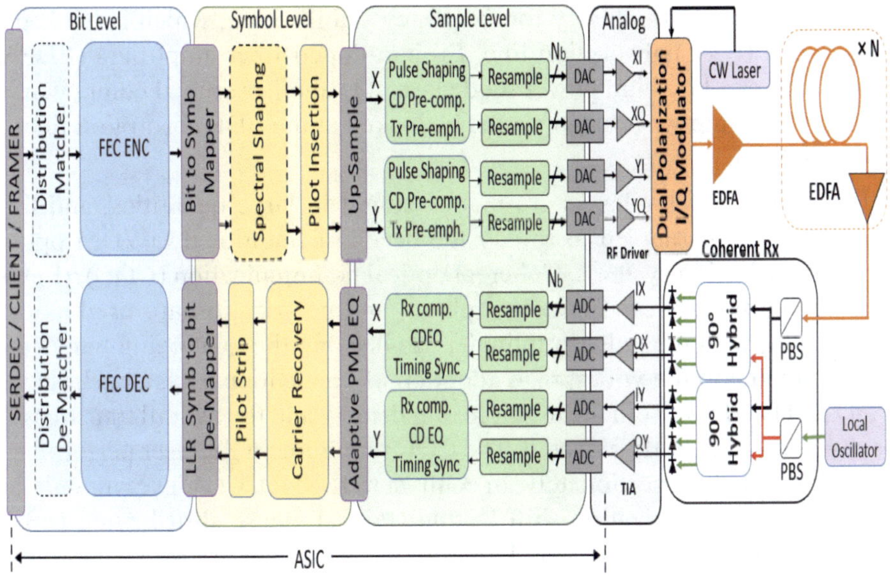

Figure 4.8. *Coherent optical communication.*

Source: Karar, A. S., Falou, A. R. E., Barakat, J. M. H., Gürkan, Z. N., & Zhong, K. (2023). Recent advances in coherent optical communications for short-reach: Phase retrieval methods. Photonics, 10(3), 308. https://doi.org/10.3390/photonics10030308, Creative Commons License.

The optical intensity has a direct relationship with the observed signal waveform. Data transfer with a coding rate more than 40 Gbit/s is currently in use. The communication system capacity is significantly increased through multiplexing.

The dense wavelength division multiplex (DWDM) is the efficient technique (Al-Saffar and Jihad, 2024). Widely used channels are spaced 50–100 GHz apart, and the wavelength spans a wide range from 1310–1550 nm (C-band) to the ultra-long wavelength of 1670 nm. As space technology advanced, optical communications between satellites and ground stations in free space became increasingly significant. Because of its high-speed response, low cost, and compatibility with optical fiber, the semiconductor laser is the essential part of optical communications. The primary cause of communication speed and capacity limitations is fiber loss and dispersion. The dispersion increases code errors and broadens optical pulses. Using a single longitudinal mode LD source, which should retain a narrow linewidth under high-frequency modulations, is the primary way to prevent the influence of dispersion. There are more LD required for various channels in the system. Thin film filters (TFF) typically supply the reference frequency for frequency stabilization, which is tracked to the acetylene (C_2H_2) absorption line. Erbium-doped fiber amplifiers (EDFA), which are pumped by high power diode lasers, is another crucial component of DWDM. The pump wavelengths must be locked at the EDF absorption lines at 980 or 1480 nm.

The semiconductor lasers are manufactured in large quantities, and their functionality is guaranteed to satisfy the needs of traditional DWDM optical fiber communication systems. Coherent optical communication (COC), (Figure 4.8) in which the optical frequency, phase, and polarization are used as data carriers along with the pulse amplitude, has become the goal being pursued in recent years due to the widespread adoption of conventional systems based on IM-DD and DWDM. Both the multiplexing density and the overall transmission capacity are significantly elevated. The COC requirement for laser performance must be increased proportionately. In contrast to IM-DD, COC uses modulation schemes such as polarization shift keying (PoSK), phase shift keying (PSK), frequency shift keying (FSK), and amplitude shift keying (ASK). Heterodyne detection must be employed in place of the direct detection utilized in IM-DD in order to demodulate the data carried by COC formats. ASK, also known as on-off keying (OOK), uses heterodyne detection at the receiver ends and has waveforms resembling intensity modulation (IM) (Cheng et al., 2025).

The received optical wave and the local oscillation (LO) correlate to produce the heterodyne signal. There are prominent benefits to correlation detection. First, LO will boost the signal strength. Secondly, the narrow linewidth correlation detection LO significantly reduces extra-band noise. Both the transmitter and the LO must use low-noise, highly coherent lasers for this reason. The correlation detection requires the employment of OPLL technology to demodulate the data carried by optical phase and/or frequency. OPLL can also effectively fix the coding impairment in long-distance transmission. DEMUX devices, such as the thin film filter (TFF), arrayed waveguide grating (AWG), and fiber Bragg grating (FBG), are utilized to implement the de-multiplexing needed in DWDM systems.

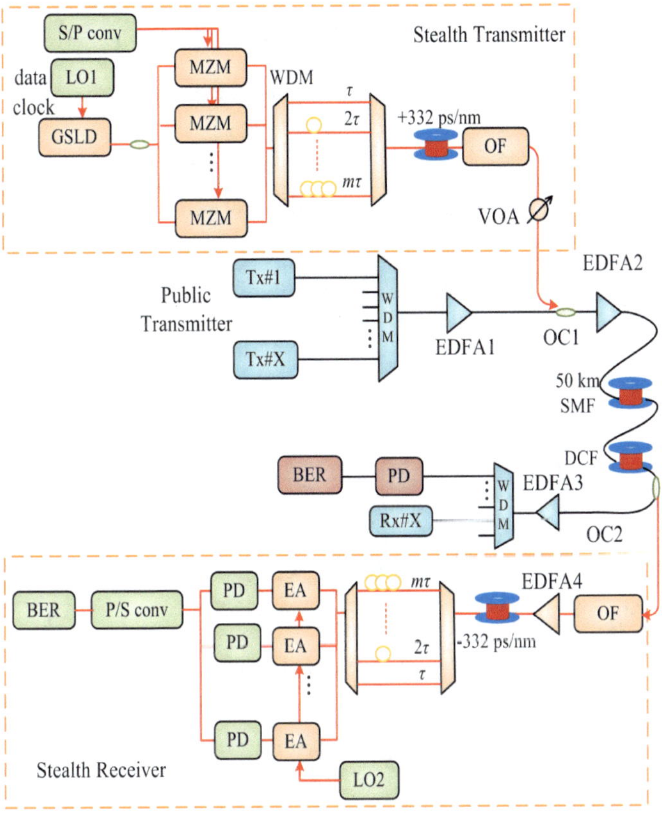

Figure 4.9. *Optical time division multiplexing.*

Source: Liu, D., Yin, Y., Cui, M., Liu, Z., & Zhu, H. (2024). Time-division multiplexed optical covert communication system based on gain-switched optical pulses. Photonics, 11(4), 376. https://doi.org/10.3390/photonics11040376, Creative Commons License.

The multiplexing density in coherence optical communications is significantly greater than in DWDM. In wireless communications, orthogonal frequency division multiplexing (OFDM) is a commonly used technology. Using the concept, coherent optical OFDM is successfully produced. The channel spacing can be lowered to the point where adjacent channels are mutually orthogonal due to correlation detection. The spectrum efficiency and multiplexing density are significantly increased. Other code formats and multiplexing techniques, like polarization multiplexing (PM) and optical time division multiplexing (OTDM), (Figure 4.9) are also being researched.

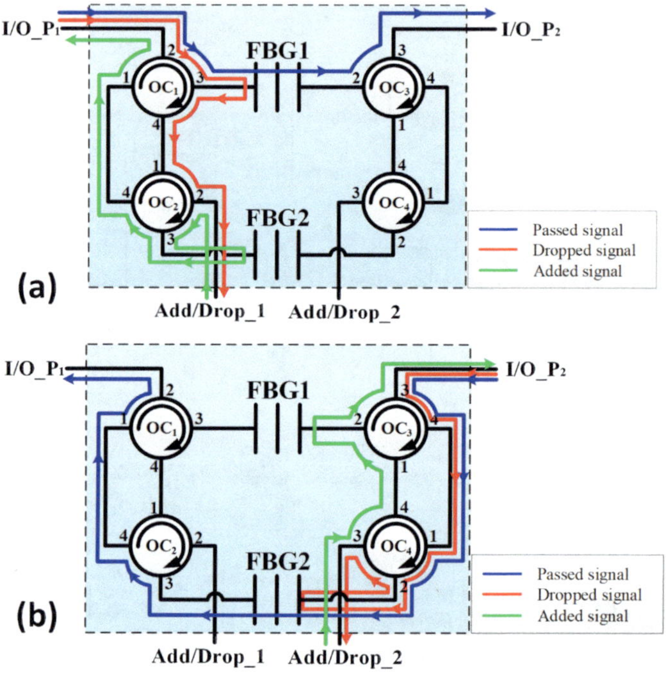

Figure 4.10. *Optical add-drop multiplexer (a and b).*

Source: Li, C. Y., Chang, C. H., & Lin, Z. G. (2021). Single-line bidirectional optical add/drop multiplexer for ring topology optical fiber networks. Sensors, 21(8), 2641. https://doi.org/10.3390/s21082641, Creative Commons License.

The most important component of the various COC systems is the narrow linewidth, extremely stable single-frequency semiconductor laser, which serves as a local oscillator in both the transmitter and the receiver. The requirements for coherence may be less stringent in COC applications than in atomic physics, but better reproducibility must be met. In normal microwave systems, the millimeter wave band and X-band microwave signals are often carried via coaxial cable or metal waveguide tubes (Faillon et al., 2008). It is now feasible to transport microwave signals via optical fiber rather than the heavy and costly waveguide tube, and to process the signals in the optical frequency domain, as the waveband of optical communications has expanded to the centimeter and millimeter range, including the generation of high-frequency signals as well as their processing and transportation. In particular, mobile communication technology has advanced to the point where practically everyone carries a cellphone. The optical fibers carry the massive amount of data. It is clear that combining wireless and optical communication is an optimal strategy, and microwave photonics is emerging as a new area of information technology. It is also known as radio over fiber (ROF), particularly in communication system local area networks (LAN) and access networks (AN), and OADM stands for optical add-drop multiplexer (Figure 4.10). Due to dispersion, optical waves carrying microwave signals would deform as they propagate through fiber, lowering the signal-to-noise ratio. Dispersion adjustment, analog-to-digital conversion, frequency down and up conversion, noise reduction, and other optical domain operations are all included in the signal processing.

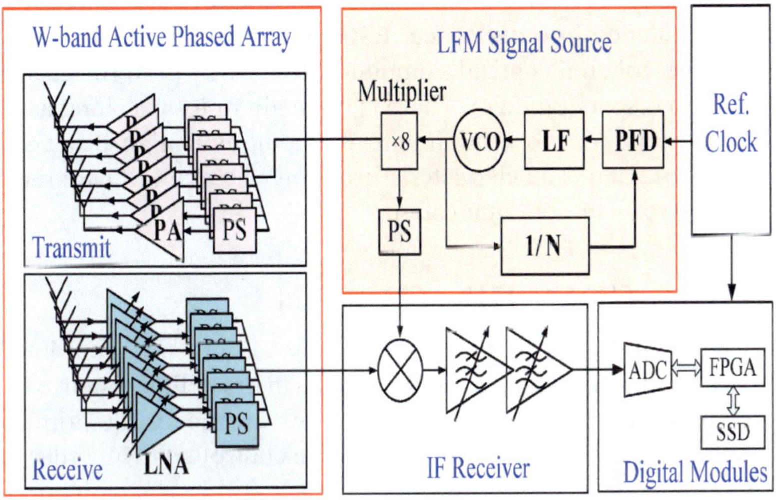

Figure 4.11. *Phased array radar.*

Source: Ding, M., Wang, X., Tang, L., Qu, J., Wang, Y., Zhou, L., & Wang, B. (2022). A W-band active phased array miniaturized scan-SAR with high resolution on multi-rotor UAVs. Remote Sensing, 14(22), 5840. https://doi.org/10.3390/rs14225840, Creative Commons License.

One of the significant uses of microwave technology is phased array radar (Figure 4.11). Its weight and volume can be significantly decreased using microwave photonics, increasing its flexibility and efficiency. By varying the phase difference between adjacent antennas, the combined beam of n antennas can be guided. Optical fibers, rather than heavy metal waveguide or coaxial cable, carry the controlling signals of microwave phases from the distant end to the antennas. The phased array antenna can provide continuous angle steering and RF-independent broadband beam steering without the beam squint effect based on the optical beamforming network. Field tests have shown a coherent radar system that is entirely photonics-based. Similar to wireless and radar, people have been using lasers for spatial communications since the early stages of their development. This technology is currently known as free-space optical communication (FSO). However, air turbulence will significantly disrupt the laser beam in the air, raising the bit error rate (BER). In such cases, the FSO in the atmosphere is essentially employed as an auxiliary tool. One of the essential components of space technology is communication between satellites and the earth; microwave communication has been the primary means of doing this. However, due to its inability to match the ever-increasing demand in terms of speed and capacity, optical communication is naturally suitable. Due to the absence of turbulence in outer space, FSO communication may demonstrate its benefits. The coherent optical communications will perform at their peak since there is no dispersion there. There is no medium loss; as long as the laser beam divergence is kept to a minimum, the communication distance can be considerably increased. The characteristics of outer space FSO are remarkable when compared to fiber communication.

4.13. SPACE COMMUNICATIONS

The spans are significantly longer. GEO is 35,786 kilometers from Earth, LEO is usually around 1000 kilometers high, and the intersatellite distance can be as much as 10 million meters. The optical beam divergence is the primary cause of the propagation loss. The divergence must be controlled to an order of micro radiance or less at such long distances. There are still issues with atmospheric turbulence and cloud and air loss for satellite-to-ground station communication. The technology of acquisition-tracking-pointing (ATP) is crucial because the laser beam is very narrow and the satellites are traveling very fast in space. A number of technologies related to optics, electronics, and mechanics are used in ATP. For communications between GEO and GS, the Doppler shift is predicted to be as high as 1/380,000 of the optical frequency (Fang et al., 2017a). The frequency shift is on the order of GHz when an 800 nm laser is utilized. The satellite velocity with respect to GS fluctuates constantly as it orbits the planet. A chirping wave is the optical signal that is received if the emitter is a single-frequency laser. Correlation detection requires the employment of a chirping-matched local oscillator; otherwise, the data will be lost. Therefore, in the outer space FSO systems, small linewidth lasers with high speed controlled tunability are essential.

CHAPTER 5

MICROCAVITY

Contents

5.1. INTRODUCTION

Almost every field of contemporary optics and photonics uses optical resonant cavities, which are made up of two or more mirrors and are a crucial component of standard lasers. Because of the reflection on the mirrors, optical energy is recirculated inside the cavities. The quality (Q) factor, which is associated with the mode lifetime and describes the light-confining ability, is one fundamental characteristic of the optical cavities (Lin et al., 2017). Another crucial component of an optical cavity is the mode volume (V), and a small V is crucial to the realization of a small integrated device. Finesse, a useful measure that considers both the mode Q factor and the resonator size, is defined as the ratio of the free spectral range to the resonance linewidth. A high-finesse microcavity with a high Q/V value is crucial for some applications.

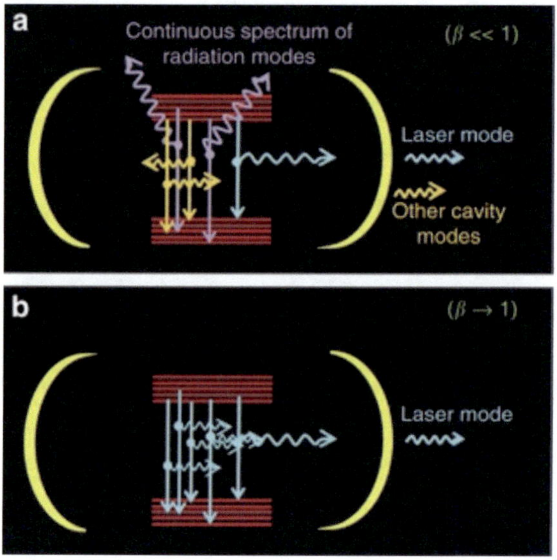

Figure 5.1. *Microcavity laser (a and b).*

Source*: Hayenga, W. E., & Khajavikhan, M. (2017). Unveiling the physics of microcavity lasers.* Light: Science & Applications, 6*(8), e17091–e17091. https://doi.org/10.1038/lsa.2017.91, Creative Commons License.*

5.2. MICROCAVITY LASER

Microcavity lasers (Figure 5.1) with a high Q/V can have a lower lasing threshold than traditional lasers (He et al., 2013). Storing optical energy in a small mode volume can significantly improve light-matter interactions. Both basic scientific research and device applications, including strong-coupling cavity quantum electrodynamics, high-sensitivity sensors, low-threshold light sources, and compact optical add-drop filters in optical communication, depend on the ability to concentrate light. A high reflectivity near unity is required to obtain high Q modes in optical cavities with a small V. This can be achieved by total internal reflection (TIR) at the dielectric boundary with a high-low refractive index contrast in whispering-gallery or WG-mode optical microcavities, or by using a periodic structure to construct a photonic forbidden band, as in vertical-cavity surface-emitting lasers and photonic crystal microcavities. The acoustical phenomena where sound waves were effectively reflected with slight diffraction and contacted the wall again at the same angle, leading them to traverse down the gallery surface, gave rise to the concept of WG mode. When the wavelengths of classical electromagnetic waves are lower than the bending radius of a reflection mirror, they can experience the same diffraction, refraction, and reflection processes as sound waves. Among the many types of optical microcavities, WG-mode microcavities are crucial for photonics integration. Many materials, including liquid droplets, polymers, glass, crystals, and semiconductors, have been examined for their ability to create WG-mode optical microcavities. The concept of WG mode was expanded to include electromagnetic waves in the RF and optical domains. Continuous TIR at the boundaries will guide the light rays, resulting in ultra-low loss for the optical modes in WG-mode microcavities. Continuous TIR light guidance is widely used in contemporary photonics and optics, including the propagating optical modes in waveguides and fibers. The light rays in WG-mode microcavities are likely to propagate along the WG surface because their incidence angles are larger than the TIR criticality at all boundaries. WG-mode optical microcavities sustain constant TIR for the confined or restricted light beams (Tobing and Dumon, 2010). For the low-power-consumption compact-size photonic devices, such as microdisk lasers, WG-mode optical microcavities with circular forms that maintain perfect rotational symmetry have attracted research interests. Circular microcavities provide ultrahigh Q factors and isotropic near and far-field patterns because the light beam contained within the cavity has a conserved incident angle above the TIR critical angle. However, effective input or output coupling is essential for a device, but it is difficult to do in the circular microcavities. In order to adjust the far-field emission patterns while maintaining the high Q factor mode, the WG microcavity can be deformed to a noncircular cavity shape.

5.3. MODE CHARACTERISTICS

High Q factors and the ability to integrate (planar) are some of the special mode characteristics of WG-mode optical microcavities where the light rays are confined by continuous TIR at the cavity boundaries (Xu et al., 2014). Due to these mode characteristics, WG-mode microcavities have found extensive use in devices. Photonic filters based on the wavelength selectivity property of WG modes, such as add-drop and all-pass filters, are a fundamental use for microcavities. A WG-mode microcavity and an optical coupler for coupling light into or out of the WG modes are components of a typical microcavity-based filter. The filtering response of an optical filter depends on the coupling coefficients for the WG modes as well as the intrinsic mode Q factors. For the majority of filtering applications, the high-order transverse WG modes are undesirable due to their stronger coupling and lower Q factors. Microrings are used to achieve high-performance optical filters and suppress high-order transverse modes. Since the modes are constrained by the continuous TIR at the outer boundaries, microrings and microdisks have comparable mode characteristics. Two evanescent-field coupled waveguides and a microring cavity are common components of an on-chip all-optical four-port add-drop filter. Pass-band filtering is demonstrated via the add-drop filter when the on-resonance light drops through the drop port. With its small size, narrow band, and wide free spectrum, microring-based filters and related active devices have been extensively researched in silicon photonics for on-chip optical connection applications. Using a higher-order filter structure with several connected microrings can further enhance the pass-band properties. Cascaded microring-based matrix switches for on-chip optical networks have been produced, based on the microring photonic filters (Poon et al., 2009). Passive networks with fixed wavelength assignment and switching networks with resonant wavelength tuning by carrier injection or thermal effect are two types of networks-on-chip. It was demonstrated that silicon-based microring electro-optic modulators may alter the microring cavity resonant wavelength by injecting or depleting carriers using the structure of an all-pass microring filter. Photonic sensing is another significant use for WG optical microcavities. The applications or uses of WG microcavities in biosensing and chemistry have been thoroughly studied. Ultrasensitive and label-free detection can be accomplished with the aid of strong light-matter interactions and high optical energy intensity in optical microcavities with a high Q/V value. Measuring the spectrum variations of a WG mode in response to environmental changes, such as the refractive index shift of media or nanoparticles onto the cavity surface, is the sensing principle. The main characteristic is the WG mode's high evanescent field, which extends from the microcavity surface and propagates along it, creating a strong interaction between the internal field and the exterior. For sensing, there will be a wavelength shift and/or splitting in the recorded transmission spectra.

Although the resonance shift in a microcavity is a simple detection method, external noise can readily interfere with it, lowering the sensing resolution (Yoshie et al., 2011). Since the two split modes experience the same noises, the mode splitting-based detection approach reduces environmental noise. The information about the particles to be measured is carried via the variation of mode splitting. The linewidth of the WG modes restricts the detection resolution for passive microcavity-based sensing. The stimulated emission in a microcavity laser-based active sensing system will reduce the linewidth, considerably improving the sensing resolution. The laser linewidth can be as small as a few Hertz for a cold cavity with a Q factor of 10^8, enabling ultrasensitive detection (Ye et al., 2024). In optics and photonics, light sources are an important area of research. The WG-mode microcavities offer a very high Q factor and a small V when compared to other optical cavities for laser applications. This allows for a variety of uses in laser physics research and the development of compact microlasers. By designing the cavity structure and coordinating the resonance wavelength with the emission wavelength of the active material, the optical density of states in WG-mode microcavities can be adjusted. The optical microcavities can significantly increase the Purcell factor. A semiconductor quantum dot is a quasi-atom gain material; in the high Q microcavity, an ultrasmall V can improve the coupling between the quantum dot and the optical mode. Investigations into cavity quantum electrodynamics and single photon sources are made easier by the strong coupling of a single GaAs quantum dot to a microdisk WG mode. Conventional low-threshold semiconductor microlasers can also be produced due to a high Q factor and small V. A WG mode strong Q factor ensures a low threshold current density, and a lower V makes the microlaser to achieve low power consumption (Xu et al., 2024). An InGaAsP microdisk laser operating at ambient temperature was able to achieve continuous wave lasing with a 40 μA threshold. However, despite a low-lasing threshold, low-output power and isotropic emission to free space were caused by the confinement of the mode light beam and the rotational symmetry of a circular microcavity. For the majority of WG microcavity laser applications, this is an important issue. One conventional method for coupling light out of circular microcavity lasers is evanescent wave coupling of a waveguide; however, this method has very high requirements for fabrication processing technology and parameter regulation. The output optical power can be decreased with a slight change in the coupling gap. The output optical power in the waveguide will become unstable due to the conflict between the clockwise and counterclockwise modes in the circular microcavities.

5.4. DEFORMED MICROCAVITY

Different deformed microcavities have been proposed and demonstrated to provide directional lasing emission. For deformed microcavity lasers, directional emission or even unidirectional emission with a low divergence angle in free space was accomplished by constructing the cavity geometries, maintaining high-Q WG modes for low-threshold lasing. Waveguide-coupled output is necessary for the application to on-chip photonic integration, and the directional or unidirectional emission of asymmetric microcavities is still restricted to free space. Because the WG modes disperse unevenly along the cavity boundaries, regular-polygonal-shaped microcavities have different mode characteristics. It is possible to realize a waveguide-coupled microcavity laser without significant disturbance to the associated high Q WG mode by using a waveguide that connects directly to the position with a weak mode field. In particular, the square microcavities with integrated internal dynamics can be solved, which makes them a dependable way to demonstrate waveguide-coupled unidirectional-emission semiconductor microlasers. Many studies have been conducted on spherical optical microcavities of liquid droplets and silica, which can have perfect microspheres because of the surface tension of the liquid and fused silica. Cavity quantum electrodynamics with changed spontaneous and stimulated emission spectra were investigated using liquid droplet microcavities as a basis, and fluorescent dyes were used to observe an ultralow threshold of nonlinear optical processes.

The dye-lasing spectra from liquid droplet optical microcavities were observed, and the effects of droplet deformation on the resonance frequencies and Q factors were examined both theoretically and experimentally using first-order perturbation theory. High-Q silica microspheres were created by joining the ends of silica fibers using the CO_2 laser fusion technique (Zhang et al., 2017). Low-threshold microlasers based on silica microspheres with doped irons were achieved, and WG modes with Q values as high as 10^9 were detected. For possible applications in integrated lab-on-a-chip systems, droplet microlasers were studied as compatible coherent light sources. Droplet-based microlasers can be made using a variety of active media, including live bacteria, in a microfluidic chip. Oil doped with a dye gain medium was injected into living cells as luminescent probes to investigate intracellular droplet microlasers. Using wafer-scale processing, ultra-high-Q microcavities can also be created on a silicon wafer as a microcavity on-a-chip that can be used in photonic integrated circuits. Using lithography and etching techniques, silica circular patterns were first converted from the photoresist layer to the thermally oxidized surface layer of a silicon wafer. Silicon was then selectively removed using silica disks acting as an etching mask. In order to prevent leakage into a high-index silicon substrate, silica microdisks on a silicon post were created using a vertical optical confinement using air. Under ideal circumstances, the WG modes with Q-factors as high as 3×10^6 were measured for these microdisks (Boriskina et al., 2006).

5.5. Q-FACTORS

The Q-factors were limited by scattering loss due to disk roughness brought on by lithography (Figure 5.2) and etching, since the mode-field distributions were situated close to the disk exterior (Yakuhina et al., 2021).

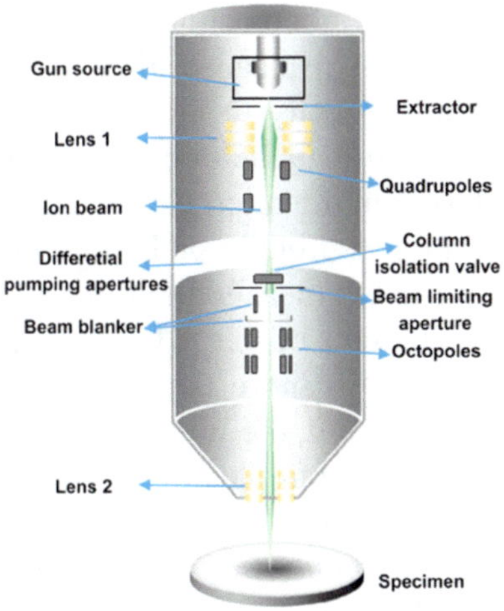

Figure 5.2. *Lithography.*

Source: Basu, P., Verma, J., Abhinav, V., Ratnesh, R. K., Singla, Y. K., & Kumar, V. (2025). Advancements in lithography techniques and emerging molecular strategies for nanostructure fabrication. International Journal of Molecular Sciences, 26(7), 3027. https://doi.org/10.3390/ijms26073027, Creative Commons License.

By smoothing the microdisk surfaces without compromising the silicon post below, a reflow method was used to further boost mode Q factors for the silica microdisks (Figure 5.3) under surface-normal CO_2 laser irradiation.

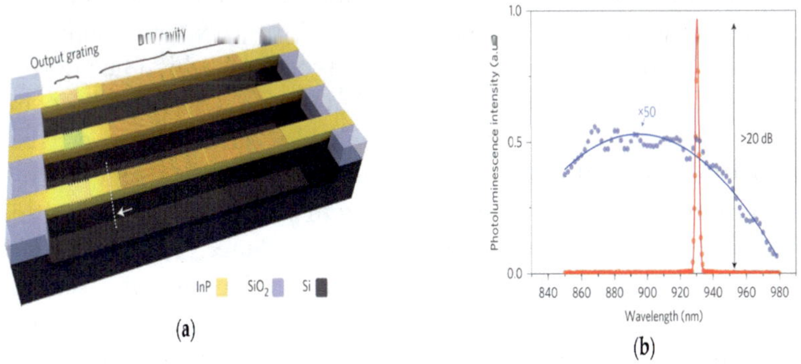

Figure 5.3. *Silica microdisk (a and b).*

Source: Zhang, J., Shankar, A. G., & Wang, X. (2024). On-chip lasers for silicon photonics. Photonics, 11(3), 212. https://doi.org/10.3390/photonics11030212, Creative Commons License.

Silica microtoroid formation on a silicon chip can result from the reflow process under laser irradiation, which can also cause silica to melt and collapse near the disk periphery. Since a loaded cavity Q factor is evaluated using a coupled waveguide, achieving an ultra-high Q factor based on linewidth measurement is difficult. Weak back scattering in the microcavity causes the WG mode to split into doublets, and thermal effects brought on by input optical power distort the resonance peak. Cavity ring-down measurement for a microtoroid cavity mode yielded an inherent cavity Q factor of 4.3×10^8 (Kippenberg et al., 2004). For microtoroids with major tori-radii more than 15 μm and observed Q values greater than 10^8 in the 1550 nm band, the WG mode loss is insignificant. Raman and Kerr nonlinearities are two examples of optical nonlinear processes that are particularly suited for research in ultra-high-Q microcavities. For ultra-low-threshold fiber-compatible Raman lasers and parameter oscillators, high-mode field intensity can be stored inside an ultra-high-Q microcavity under fiber evanescent coupling with a low-input power at resonant frequency. Using a high Q microresonator and a cascaded four-wave mixing technique, optical frequency combs were achieved. Chemically etched wedge resonators on-a-chip were created by standard semiconductor processing; they surpassed microtoroids with a Q factor of 875 million (Lee et al., 2012). By extending the chemically etched time to generate wedge profiles for the resonator perimeter and curing the roughness of photoresist patterns, the post-exposure bake approach increased the smoothness of wedge resonators. The wedge resonators are easily controllable in size and can be combined with other photonic devices as they do not require the reflow process of laser irradiation. Multiple-layer semiconductor materials, such as AlGaAs/GaAs and InGaAsP/InP material systems, which may produce a multilayer slab waveguide, are typically used to grow semiconductor lasers. The simple scenario is the multilayer slab waveguide, which has normal mode characteristics for both guided and radiated modes. Under boundary conditions applied to the mode fields at the waveguide dielectric interfaces, the modes are solutions of Maxwell equations. We can simulate z-direction propagating modes under uniform distribution conditions in y-direction for a multiple-layer slab waveguide with interfaces parallel to the y-z plane. This corresponds to the transverse mode distribution for edge-emitting semiconductor lasers. With the z-direction component of the electric field and magnetic field taken to be zero, the z-direction propagating modes under the y-direction uniform condition can be divided into transverse-electric (TE) and transverse-magnetic (TM) modes. There are several distinct longitudinal modes that are appropriate for measuring gain, and there are no minor peaks of higher-order transverse modes. The mode gain spectrum can be determined using the Hakki-Paoli method, which is based on the peak-valley ratio of the transmission spectrum of the Fabry-Perot cavity and the peak-valley ratio of the amplified spontaneous emission spectrum below the threshold. However, the resolution of the measuring instrument/tool affects the measured amplified spontaneous emission spectra, causing the measured spectrum peaks to drop and its troughs to grow.

5.6. MODE GAIN

Based on the integrations of the product of an amplified spontaneous emission spectra and a phase function across a single mode interval, the Fourier series expansion approach is suggested as a means of measuring mode gain. Compared to the Hakki-Paoli approach, the Fourier series expansion method is less susceptible to noise since it uses the entire resonance pattern to evaluate mode gain.

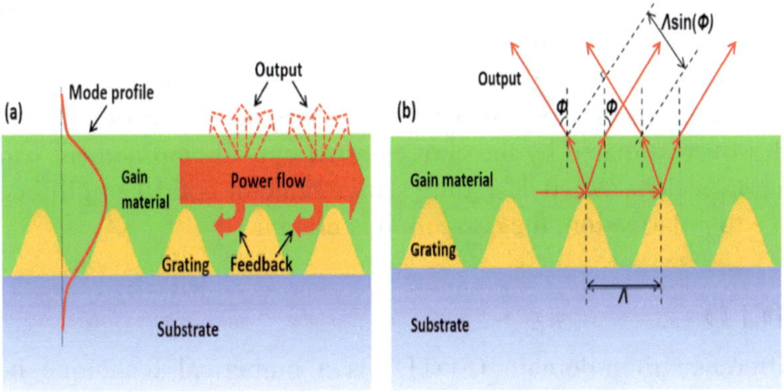

Figure 5.4. *Distributed feedback (DFB) laser (a and b).*

Source: Zhou, P., Niu, L., Hayat, A., Cao, F., Zhai, T., & Zhang, X. (2019). Operating characteristics of high-order distributed feedback polymer lasers. Polymers, 11(2), 258. https://doi.org/10.3390/polym11020258, Creative Commons License.

By compensating for the influence of the low-resolution measuring system response function, an accuracy gain spectrum can be derived. In order to achieve single-longitudinal-mode functioning, VCSEL small cavity ensures a huge gain difference between various longitudinal modes due to its large longitudinal mode interval. Based on the mode gain difference, edge-emitting Fabry-Perot cavity semiconductor lasers are challenging to achieve dynamic single-longitudinal-mode operation. These lasers typically have a cavity length of around 100 μm. To obtain single-longitudinal-mode functioning edge-emitting lasers, advanced mode selection devices are needed, such as cavity mirrors made of distributed Bragg reflectors (DBRs) (Gryga et al., 2022). For the longitudinal-mode selection, the DBRs, which are made up of gratings with a modest diffractive index, can have a narrow band high reflectivity spectrum. Since VCSELs have a huge refractive index difference and a relatively short cavity, they can actually be thought of as DBR lasers with a wide band high reflectivity spectrum. Distributed feedback (DFB) lasers (Figure 5.4) are popular single-mode semiconductor lasers with a cavity with gain and grating, in contrast to DBR lasers with passive DBR zones. We can model a DFB laser and derive the eigenvalue equation of the DFB lasers by using the grating as a step variation of the refractive index. The widely used semiconductor lasers are edge-emitting lasers that meet the Fabry-Perot cavity phase requirement and have longitudinal modes. The active and passive segments of a multiple-layer waveguide for edge-emitting semiconductor lasers limit the resonant modes in the direction of material growth (Tong, 2014). One important element for determining mode gain and creating laser structures is the optical confinement factor. The ratio of the z-component power flow in the active region to the total power flow is known as the standard optical confinement factor. The research findings show that the mode gain of TE modes will be underestimated and that of TM modes overestimated by the conventional optical confinement parameters based on power flow. When designing semiconductor optical amplifiers that are insensitive to polarization, the correction is crucial.

5.7. FDTD

Finite-difference time-domain (FDTD) is a numerical technique based on Maxwell equation that uses finite difference rather than differential to deal with electromagnetic fields in the time domain (Yee and Chen, 1997). Taflove and Hagness developed it after Kane Yee proposed it. The FDTD approach has grown into a popular numerical technique for resolving electromagnetic field problems as a result of the quick advancement of computing technology. Since the FDTD method is directly derived from Maxwell equation, it has no approximation when compared to other numerical methods for simulating electromagnetic fields, such as the beam propagation method (BPM), finite-element method (FEM), etc. It may also accurately reflect the physical nature of the electromagnetic problem. The FDTD approach is more flexible than previous approaches and can easily mimic electromagnetic field issues in a variety of intricate systems. While studying optical modes in microcavities, the FDTD method is an important approach. But there are drawbacks to the FDTD approach as well. Large amounts of memory and processing time are needed to simulate a large-scale structure; these issues will be resolved progressively as computing technology advances. The FDTD method operates in the time domain, and by converting the time-domain signal into the frequency domain, the spectrum in the frequency domain can be acquired. One of the most popular techniques is the fast Fourier transform (FFT), however it usually requires extensive time series data. The time-domain signals must then be converted into the frequency domain. The Padé approximation method, which is based on the Baker algorithm, can reduce the time needed for the FDTD simulation and get a high-resolution spectrum from short time series data. The FDTD approach linked to the Padé approximation may be a useful tool for resolving electromagnetic issues in optical microcavities and the related system. Because of the limitations of computer storage and computation time, the infinite structure cannot be directly calculated while studying the electromagnetic field problem in open systems using the FDTD approach. The simulation window should be closed using a suitable absorption boundary condition. The absorption boundary is to approximate a finite space into an infinite space by absorbing the computational space propagating wave as much as feasible without producing pseudo-reflection. Creating appropriate absorbing boundary conditions has always been a key component of the FDTD method development. Mur's absorbing boundary conditions and perfectly matched layer (PML) absorption boundaries are now the effective boundary absorption conditions (Hu, 2004). Mur's absorption boundary condition will result in pseudo-reflection since it is constructed using the traveling-wave equation under the plane-wave approximation. According to Berenger PML condition, an absorbing substance is introduced on the boundary, and once an electromagnetic wave enters it without reflection, it is attenuated. PML is frequently utilized in the FDTD simulation and is an effective absorption boundary in computations. Even while the Padé approximation reduces the running time for FDTD simulation, the time-to-frequency transformation requires more time because Padé approximation often requires more computing time than DFT for the same-length time-series data. Because the transformations for the electric and magnetic field components must be performed at every location across the plane of flux distribution, the situation for the spectral response computation gets worse. Typically, just the fundamental transverse mode is supported by the bus and drop waveguides. Each bus and drop waveguide port located far from the coupling region should have the same field distributions. In order to create an electromagnetically induced transparency (EIT)-like effect, the structure parameters are changed to have the same resonance wavelength for the racetrack resonator and the inside microring with the mode number difference is an odd number. Using the Padé approximation and the FDTD approach, we compute the time delay. The input pulse in the FDTD simulation is set using the bootstrapping approach. The ratio of field intensities at the output plane center points to the input plane can be used to approximate the ratio of Poynting power flux over the output plane to the input plane.

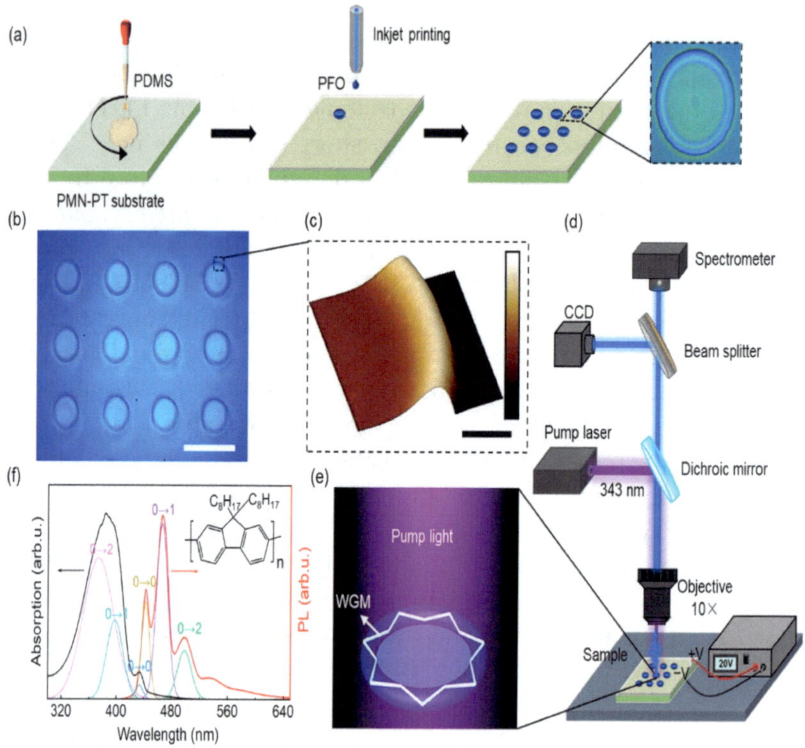

Figure 5.5. *Whispering-gallery-mode (WGM) microcavity laser (a-d).*

Source: Liu, F., Tong, J., Xu, Z., Ge, K., Ruan, J., Cui, L., & Zhai, T. (2022). Electrically tunable polymer whispering-gallery-mode laser. Materials, 15(14), 4812. https://doi.org/10.3390/ma15144812, Creative Commons License.

Semiconductor whispering-gallery-mode (WGM) microcavity laser (Figure 5.5) small footprints, high quality (Q) factors, and planar geometry make them promising energy-efficient and compact light sources for photonic integration. Due to their high-Q WGMs, WGM microcavity lasers with a circularly symmetric geometry have gathered attention from researchers. However, a significant drawback of circular microcavity lasers is that, because of their circularly rotating symmetry, their lasing emission is usually nondirectional and uniform around the cavity rim, which is unsuitable for applications. Researchers have been examining several deformed microcavities to appropriately break the cavity symmetry and preserve high-Q WGMs with effective output coupling in order to achieve directional lasing emission from WGM microcavity lasers. For directional emission, WGM microcavity lasers with smoothly deformed cavity geometries were suggested. Many asymmetric microcavity lasers, such as spiral and limacon microcavity lasers, were able to produce unidirectional lasing emissions by designing the cavity shapes. Waveguide-coupled deformed microcavity lasers for on-chip optical interconnects have been demonstrated. InP/InGaAsP disks on a pedestal with InGaAs quantum wells as the active medium under optical pumping were the first semiconductor microdisks to exhibit lasing. Researchers demonstrated room-temperature pulsed injection lasing with a submilliamp threshold using a double-disk structure. The current was injected via the top and bottom pedestals, and the top disk served as the p-contact layer. However, because the WGMs primarily disperse in the rim region, the pedestal-mounted microdisks restrict the current injection efficiency and stop heat dissipation. The electrically pumped InGaP/InGaAs microdisk laser, also known as a microcylinder laser, that is vertically confined or limited by semiconductor materials was demonstrated (Maximov et al., 2020). The weak vertical confinement causes additional radiation loss and may restrict the laser size. Since the initial demonstration of microdisk lasers, circular microdisk semiconductor lasers with various material platforms have been thoroughly studied for their potential uses in photonic integration and physics research due to their high Q WGMs. The isotropic emission along the cavity rim, however, is a drawback that makes it difficult to efficiently collect the output light from the circular microdisk lasers.

5.8. EMISSION

The circularly rotational symmetry must be broken in order to achieve directional or unidirectional emission from the WGM microcavity lasers. By scattering or refracting light out from the discontinuous boundary, local perturbation is a logical concept for achieving directed emission. With some designs, the local boundary defects might cause directional or unidirectional emission in free space by breaking the cavity symmetry. Researchers experimentally showed that a microdisk laser with a local boundary defect can enhance the emission directionality shortly after the initial demonstration of a semiconductor circular microdisk laser. A perfect microdisk laser would emit light in a symmetrical pattern. The far-field emission of a microdisk laser usually did not conserve circular symmetry and exhibited considerable edge roughness in the form of radius variations. The emission from the laser output was uncontrolled and nondirectional. Without significantly raising laser thresholds, it may be possible to adjust the direction and intensity of light output by creating patterned asymmetries in the cavity form. A linear grating across the disk circumference could be used to create the asymmetries. While the emission directionality was enhanced, these distorted microcavities lased with thresholds almost twice as high as their ideal circular disks. A double-disk structure microcavity laser was experimentally proved to couple light output from a microdisk laser and sustain a high Q value. The double-disk microcavity laser was made up of a top waveguiding disk with a notch that served as a leaking source for output coupling and a lower lasing disk with a multiple quantum well (MQW) active layer. While the nondeformed bottom lasing disk kept a high cavity Q value, the top disk lasing light coupled out horizontally. It has been shown that notch-equipped microdisk lasers may simultaneously control the lasing modes and achieve directed emission. Researchers conducted studies on the Q factor and emission pattern control of the WGMs in notched microdisk resonators. The findings demonstrated that the notched microcavity design could effectively adjust the directed light output, frequency separation, and Q factors of two symmetrical types of double-degenerate WGMs. Scientists showed unidirectional emission of a WGM elliptical microcavity quantum-cascade laser (QCL) with a wavelength scale notch at the boundary using midinfrared quantum cascade lasers as a model system. The beam divergence showed the durability of the cavity design by being insensitive to both the notch shape and the pumping current. The fundamental principle is that the notch can function as a minute scatterer, sending light rays in the direction of the elliptical disk opposite side, where the elliptical border collimates the outgoing rays into a parallel beam.

Copropagating pairs with dominant components in the CCW direction are produced by the asymmetric scattering in the spiral-shaped microcavity (Wiersig et al., 2008). By laterally connecting a waveguide to the cavity, unidirectional emission can be achieved. A spiral-ring microcavity structure was employed to increase the current injection efficiency since the WGMs primarily disperse in the rim area. For the hybrid spiral-ring microcavity laser, the power from the CCW direction is over five times more than that from the CW direction. The hybrid spiral-ring laser achieves unidirectional emission from the CCW direction and eliminates the output power instability present in the hybrid microdisk and ring lasers. Unidirectionally emitting in-plane waveguide-coupled microcavity lasers must be a viable option for on-chip optical interconnect applications. The spiral microcavity laser, which has the primary advantage of direct gapless coupling from a distinctive notch that is only a fraction of the microcavity radius to an integrated waveguide, is one competitor for such a waveguide-integrated on-chip light source. The asymmetric scattering caused by the notch can be eliminated by the coupled waveguide, guaranteeing an efficient output through the waveguide. A high waveguide-coupling efficiency through the spiral notch is suggested through the simulation findings. The ratio of the energy flow via the output waveguide to the total energy flux coupled out of the cavity is known as the waveguide-coupling efficiency. For the simulated WG-like modes, the simulated waveguide-coupling effectiveness is more than 70% (Song et al.,

2009). The field-amplitude distributions inside the waveguide indicate several high-order transverse modes, indicating a multimode waveguide. Ring-shaped p-contacts are employed on top of the spiral rim region to spatially overlap the current injection with the spiral microcavity high-Q modes. For the AlGaInAs/InP waveguide-coupled spiral microcavity lasers with a disk radius of 30 μm, room-temperature continuous-wave electrically injected lasing was achieved. The lasing threshold can be effectively lowered by the ring-shaped current injection. Spiral microcavity lasers were shown to be capable of high-speed direct modulation. Following the initial demonstration of microdisk lasers, research was done on deformed microcavity lasers with smoothly deformed boundaries in addition to deformed microcavities with discontinuous boundaries in order to achieve directed or unidirectional emission. The problem between the Q factor and directionality is crucial for deformed microcavities because the deformation is usually accompanied by the Q spoiling. Waveguide coupling (Figure 5.6) is required for the on-chip implementation of disordered microcavity lasers.

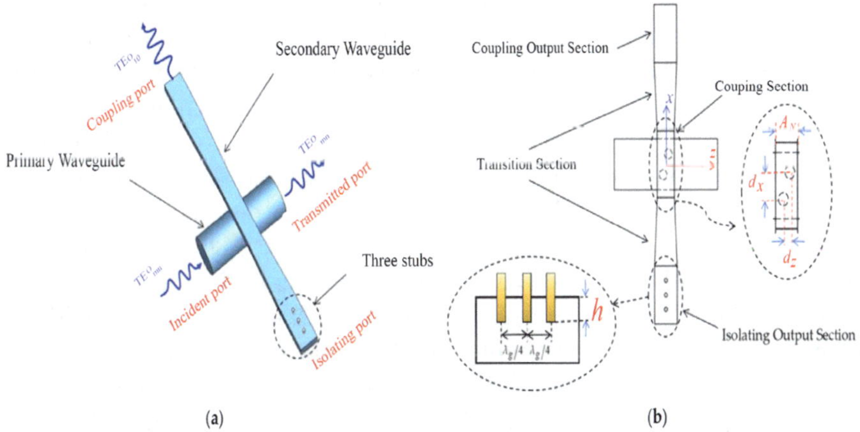

Figure 5.6. *Waveguide coupling (a and b).*

Source: Zhu, T., Fu, W., Lu, D., Pan, Y., Li, C., He, L., Sun, H., & Yan, Y. (2024). A compact circular waveguide directional coupler for high-order mode vacuum electronic devices. Electronics, 13(3), 633. https://doi.org/10.3390/electronics13030633, Creative Commons License.

5.9. MODES

Accurate alignment is necessary for the conventional wave coupling. Since the waveguide also acts as a perturbation that significantly alters the mode-field distributions, it is possible to get efficient output with high Q reconstructed modes by directly attaching a waveguide to a nondeformed circular microcavity. This is because of the mode coupling between various radial-order modes. A waveguide coupled to the position with a weak mode field can couple the laser light out without causing severe Q spoiling because the mode fields in polygonal microcavities spread unevenly along the cavity boundary. High Q modes are usually restricted to the regular islands surrounding the stable periodic orbits in the moderately deformed microcavities with mixed-phase space (Cao and Wiersig, 2015). The light rays of these modes are completely contained by total internal reflection in classical ray optics, meaning they are unable to escape. Because of their small size and high quality (Q) factors, semiconductor whispering-gallery-mode (WGM) microdisk lasers can achieve low-threshold operation. However, the circular microdisk rotational symmetries severely restrict their directional emission and output power. It has been shown that a variety of microdisk lasers with evanescently coupled waveguides or deformed geometries can achieve effective directional emission. The waveguide output was achieved by directly connecting a waveguide to the position with weak mode fields, and polygonal microcavities can support high QWGMs with nonuniform mode-field patterns throughout the cavity perimeter. In a circular microdisk, the mode fields of WGMs disperse evenly around the edge. Since the light beams of the WGMs would approach the output waveguide exit window after a period causes a significant leakage loss, we may anticipate that the mode Q factors should drastically destroy when a waveguide is connected directly to the cavity boundary. The circular symmetry of the microdisk is broken by the connected waveguide and mode coupling between various radial-order WGMs with close wavelengths results in high Q-coupled modes with field distributions localized along geometric structures that stop leakage through the connected waveguide. It has been experimentally shown that semiconductor microdisk lasers that couple one or more waveguides can produce either directed or unidirectional emission. In a nondeformed circular microdisk, the mode coupling between WGMs with differing angular mode numbers is restricted. The mode coupling between two modes with close wavelengths occurs because of the symmetry break for the microdisk with an output waveguide that is radially coupled to the cavity boundary. Two new coupled modes will be produced by the mode coupling, in contrast to the mode superposition. The finite-difference time-domain (FDTD) approach is used to simulate a 4.5-μm diameter microdisk with a refractive index of 3.2 and an air-enclosed 0.6-μm wide output waveguide. The modes of various symmetries are independently simulated using symmetric or antisymmetric excitation sources. The FDTD computation window is terminated using the perfect matched layer (PML)-absorbing boundary condition as the boundaries. Different coupling phenomena with various coupling coefficients can be observed by altering the mode frequencies. The external mode coupling results in splits in both the real and imaginary parts of the mode frequency if the uncoupled two modes have exactly the same mode frequency (□1 = □2). This implies that the two coupled modes will have reduced and enhanced mode Q factors, respectively. Long-lived scar-like modes in the optical microcavities have been demonstrated to be produced by external coupling. The waveguide output coupling and scattering centered in the waveguide-connecting area are the main losses in the waveguide-connected microdisk. External mode coupling results from interference between two neighboring modes in the loss area; the strength of the coupling is proportional to the two mode losses. The mode coupling effect will be evident when the coupling strength is equal to the frequency difference between the two WGMs. In the loss region, destructive interference is associated with high Q-coupled modes.

High radial-order WGMs can have high Q factors in a relatively large microdisk, and that means there will be many high Q modes in a single free spectral range (FSR) (Chiasera et al., 2010). The FDTD simulation and Padé approximation for the TE modes yield the intensity spectrum containing one FSR for a circular microdisk with a radius of 10 μm, a refractive index of 3.2, and an output waveguide that is 2 μm wide. The FDTD simulation can produce single mode-field patterns for the high Q modes by selecting a narrow-band excitation source centered at the resonant frequency. Waveguide-connected microdisks can have high Q coupled modes with the field patterns created by the superposition of two WGMs with close wavelengths. An AlGaInAs/InP multiple quantum well (MQW) epitaxial wafer was used to create waveguide-connected microdisk lasers. We use the FDTD technique to simulate the mode Q factor and output coupling efficiency vs. output waveguide width by connecting an output waveguide to one of the ETR vertices. First, we excite all of the TE modes under consideration using pulse generators of broad frequency range. Next, we use the Padé approximation approach to determine the intensity spectra from the FDTD output. Following the acquisition of the mode frequencies and Q-factors, we excite the associated mode field distribution by using an exciting source to cover a small frequency range centered at the obtained mode frequency. The energy flow confined in the output waveguide divided by the total energy flux around the entire resonator, which is determined by integrating the Poynting vector over a period, is used to estimate the output efficiency based on the mode field distribution. The symmetric and antisymmetric modes with respect to the output waveguide midline and the ETR are simulated using symmetric and antisymmetric exciting sources. First, a hard mask for dry etching was applied to the as-grown InP/GaInAsP laser wafer using plasma-enhanced chemical vapor deposition (PECVD) (Figure 5.7) to manufacture an 800-nm SiO_2. After that, conventional photolithography and ICP etching methods are used to transfer the

ETR patterns onto the SiO_2 layer. The ETR geometry for the ICP process to etch InP and GaInAsP using BCl_3-Cl_2 gas mixtures is defined by the patterned SiO_2 hard masks. The vertical penetration depth of the optical field of the confined modes is less than the ICP etching depth, which is around 4 μm. Following ICP etching, a chemical etching procedure employing a Br_2 + Hbr+H_2O solution is used to eliminate reactants from the side walls created during the ICP etching process and enhance their smoothness. After that, diluted HF solution is used to remove the remaining SiO_2 hard masks.

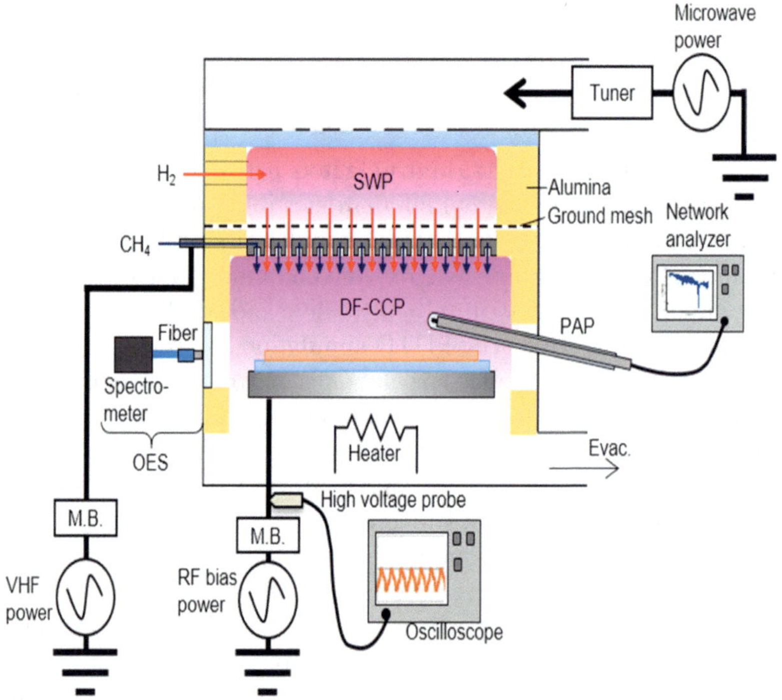

Figure 5.7. *Plasma-enhanced chemical vapor deposition.*

Source: Sugiura, H., Kondo, H., Tsutsumi, T., Ishikawa, K., & Hori, M. (2019). Effects of ion bombardment energy flux on chemical compositions and structures of hydrogenated amorphous carbon films grown by a radical-injection plasma-enhanced chemical vapor deposition. C, 5(1), 8. https://doi.org/10.3390/c5010008, Creative Commons License.

Each ETR pattern is covered with a 300 nm SiO_2 insulating layer, and a contact window is created using a wet etching technique. Standard metal deposition is used to create a top Ti-Pt-Au p-contact, and the wafer is lapped down to a thickness of roughly 100 μm (Yan et al., 2022). For the n-type contacts, Au/Ge/Ni/Au metallization is employed. High-Q-coupled WG-like modes would arise from the mode coupling between the modes with even longitude mode numbers. An output waveguide can be directly connected to the position with weak field distribution due to the WG-like mode sinusoidal-modulated envelope along the sidewall. Electrically injected lasing is achieved at room temperature using InP-based waveguide-coupled directional-emission square microcavity lasers. Because various transverse modes have different field distributions, dual-mode lasing with tunable intervals is proven by modulating the lasing modes with a spatial current injection. The dual-mode microlaser is used to produce a tunable microwave signal. OFCs and sub-THz waves are produced using dual-mode square microcavity lasers. A single-mode square microlaser with a broad continuous tunable wavelength range is demonstrated by appropriately engineering the waveguide width to regulate the mode Q factor difference between various longitudinal modes. In order to improve mode confinement and extend the dual-mode interval from the tens of GHz to the THz range, deformed square microcavities with circular arcs in place of the flat sidewalls are suggested. The application of optical sensors may benefit from the ultra-high mode Q factors for the deformed square microcavity.

CHAPTER 6

SEMICONDUCTOR DEVICES

Contents

6.1. INTRODUCTION

We can use stimulated emission from interband transitions through semiconductor lasers that are solid-state lasers based on direct bandgap semiconductor materials. The high carrier density in the conduction band causes these transitions. All of the electrons are in the valence band when there is no external stimulus, such as an optical pump. Some of these electrons can be forced high into the conduction band by an external optical pump beam with photon energy above the semiconductor band gap energy. The electrons return to the metastable state, which is the lowest energy level in the conduction band, from these high-energy excited states. The holes that are created in the valence band simultaneously migrate to the top of the valence band. Photons with an energy close to the bandgap energy can then be released when electrons near the bottom of the conduction band recombine with these holes. An external pump that supplies photons with the appropriate energy levels enhances this process. The conduction band electrons in the holes of semiconductors with indirect band gaps, like silicon, acquire different wavenumbers. Because momentum is not conserved, optical transitions are therefore impossible.

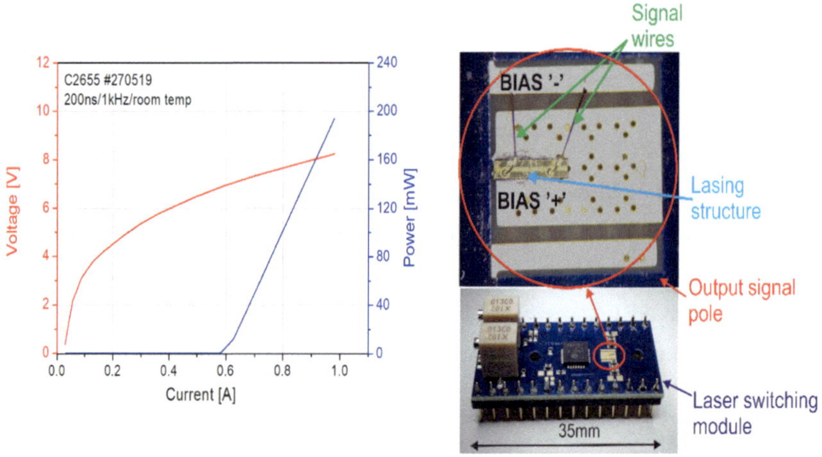

(a) (b)

Figure 6.1. *Quantum cascade laser (and b).*

Source: Mikołajczyk, J., & Szabra, D. (2021). Integrated IR modulator with a quantum cascade laser. Applied Sciences, 11(14), 6457. https://doi.org/10.3390/app11146457, Creative Commons License.

6.2. QUANTUM CASCADE LASERS

Quantum cascade lasers (Figure 6.1) and optically pumped semi-conductor lasers are also available, but optical diode lasers make up the majority of semiconductor lasers. Electrical pulses are used to pump diode lasers at the np junction. Quantum cascade lasers use intraband transitions, while optically pumped lasers employ an external optical pump. Among the materials used in direct bandgap semiconductor lasers are Aluminum arsenide (AlAs), Aluminum gallium arsenide (AlGaAs), GaP (Gallium phosphide), InGaP (Indium gallium phosphide), GaN, or Gallium nitride, Indium Gallium arsenide (InGaAs), Indium phosphide (InP), Indium gallium Nirtride arsenide (InGaNAs), and Indium Gallium phosphide (InGaP) (Hardtdegen and Mikulics, 2019). Various compositions with different bandgap energies permit diverse emission wavelengths since the photon energy of a laser diode is near the bandgap energy. The bandgap energy of ternary and quaternary semiconductor compounds can be continually changed within a range. An increase in aluminum content raises the bandgap energy in AlGaAs. The majority of semiconductor lasers function in the near-infrared spectrum; some, such laser pointers made with GaInP, produce red light, while others use gallium nitrides to produce blue or violet light. Lead selenide (PbSe) lasers and quantum cascade lasers are a few examples of mid-infrared emission lasers. Research is being done on organic semiconductor compounds for application in semiconductor lasers. Electrical pumping is challenging to achieve high efficiency for a variety of reasons. There are many different types of semiconductor lasers available, each with a distinct application. High-quality low milliwatt (0.5 W) laser diodes with small edge are used in optical fiber communications, CD players, and laser pointers. A laser diode serves as the gain medium of a longer laser cavity in external cavity diode lasers. They have a narrow emission linewidth and are frequently wavelength tunable.

6.3. POWER

For the production of ultra-short pulses, mode locking is also possible for both external-cavity and monolithic low-power levels. Low beam quality, low output power beams are produced by broad area laser diodes. High power diode bars are a collection of broad-area emitters that provide output of low quality and high power. High power can be produced via high power diode bars. Low power, high beam quality laser radiation is emitted perpendicular to the wafer by surface emitting lasers (VCSELs). Even when operating in mode lock, optically pumped vertical cavity surface emitting lasers (VECSELs) produce multi-watt output with good beam quality. Similar performance is expected with electrically pumped photonic crystal surface emitting lasers. In contrast to interband transitions, quantum cascade lasers work on intraband transitions and typically emit mid-infrared beams with a teraHertz frequency. Optical fiber communications use semiconductor lasers with wavelengths between 1.3 and 1.6 microns (Suematsu, 1983). These make use of the quaternary compound $In_{1-x}GaxAsyP_{1-y}$, which is generated via molecular beam epitaxy in a layer form on InP substrates. To maintain a defined lattice structure and prevent defects from forming at the interfaces between any two layers with differing bandgaps, the lattice constants of each layer must match the lattice constant of InP. The semiconductor material that emits light is the central core layer or active layer. The bandgaps of the cladding layers are larger than those of the active layer. Electrons and holes are restricted to the active layer by the difference in bandgap between the two semiconductors. The active layer functions as a planar waveguide with a slightly higher refractive index than the surrounding cladding layers. By adjusting the active-layer thickness, the number of modes can be adjusted. Both the injected carriers like electrons and holes and the light produced by electron-hole recombination inside the active layer are contained by this heterostructure design. Because of their higher bandgap, the two cladding layers are also transparent to the light that is emitted, creating a low-loss structure. Because of these characteristics, semiconductor lasers can be applied to a wide range of applications. Population inversion and optical gain are observed in the active region when the injected carrier density in the active layer surpasses a threshold. A semiconductor laser generates light in multiple longitudinal modes of the cavity. The gain spectrum of semiconductor lasers is broader enough to allow for the simultaneous experiencing of gain by several longitudinal modes of the FP cavity. The dominant mode is closest to the gain peak. Since the other mode gain is always less than the main mode, they should not exceed the threshold. The difference is very less that a significant portion of the laser power is shared by the main mode and one or two surrounding modes on either side of it. The multimode characteristic of a semiconductor laser frequently restricts the bit rate of lightwave systems operating close to 1.55μ since each mode propagates inside the fiber at a different speed due to group velocity dispersion. Lasers can be made to oscillate in a single longitudinal mode to overcome this problem.

6.4. VECSEL

Another semiconductor laser is the vertical cavity surface emitting laser (VECSEL), which consists of a laser resonator and a semiconductor integrated circuit that emits surface light. A VECSEL may provide high beam quality, high power, and diffraction-limited optical output, in contrast to other kinds of semiconductor lasers. In contrast to gas lasers and doped insulator solid state lasers, a VECSEL output wavelength can be tunable. The active light generating region of a VECSEL is made up of many quantum wells and one or more Bragg mirrors, with a total thickness of a few microns. The entire device is installed on a heat sink, and the laser is made on a semiconductor substrate. An external mirror placed at a distance of a few millimeters to a few centimeters is used to construct the laser resonator. The resonator length cannot be less than the intracavity beam Rayleigh length in order to achieve good beam quality output. The laser mode is controlled by this external resonator. Other optical components and flat curved mirrors may be folded into the external resonator. These extra optical components consist of an optical filter that operates at a single frequency, an intracavity frequency tuning crystal that is nonlinear, and a passive mode locking saturable absorber. It is possible to pump a VECSEL optically or electrically. A ring electrode is utilized to make contact with the active area in electrically pumped VEC-SEL. Because it is challenging to pump large areas uniformly and prevent a weakly pumped region at the center of the active area, this technology restricts the useable active area and the output power. The higher output power that electrically pumped VECSELs can produce is 1 W. The drawbacks of electrical pumping are avoided by using the optical pumping technique. Uniform illumination of large active regions is made possible by optical pumping. There is no requirement for specifically doped regions to transport current.

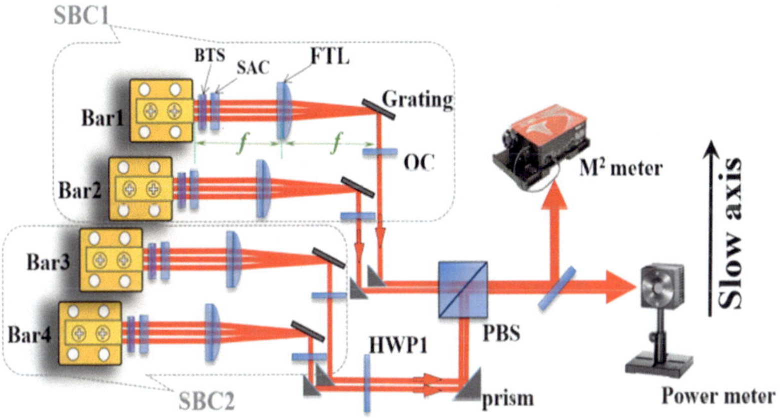

Figure 6.2. *Diode bar.*

Source: Zhao, R., Zhao, Y., Meng, B., & Tong, C. (2025). High-brightness 1940 nm gallium antimonide diode lasers with external-cavity spectral and polarization beam combining. Photonics, 12(6), 594. https://doi.org/10.3390/photonics12060594, Creative Commons License.

A diode bar (Figure 6.2) or high brightness broad area laser diode provides the pump light. The beam quality of the pump light is irrelevant because the semiconductor gain structure has a very short absorption length. If the beam quality is poor, it only necessitates using a strongly converging pump beam, which takes up more room and could make it more challenging to arrange the intracavity elements. More watts of output power by a diode bar solves this issue. When the mode area is large, diffraction limited output requires an external resonator. Even with optical pumping, VCSELs are not appropriate for high powers with good beam quality (Michalzik and Ebeling, 2003). If only the quantum wells are present, the input optical power absorption is minimal due to their thinness. In order to overcome this problem, incoming optical pump radiation is absorbed by a gain structure made up of spacer layers positioned in between the quantum wells. Since the quantum wells have a narrower band gap than the spacer layers, the carriers produced in these layers can be effectively transported to them. The bandgaps of the spacer and quantum well materials must be widely separated for efficient carrier transfer, which necessitates a pump wavelength that is significantly shorter than the laser wavelength. However, direct pumping of the quantum wells is a suggested in well pumping. As in a solid-state thin disk laser, a multipass pumping scheme can be used to provide efficient input pump power absorption. This VECSEL is challenging to fabricate because the pump radiation optical spectrum needs to meet tolerances.

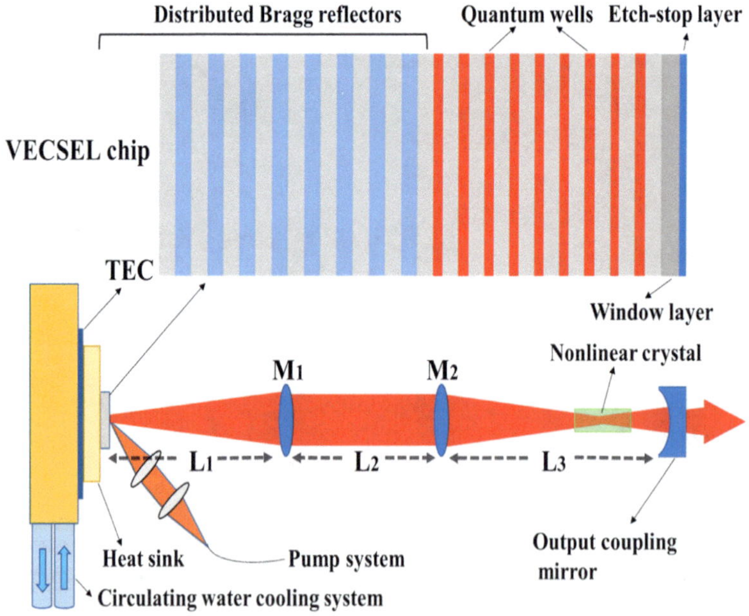

Figure 6.3. *VECSEL quantum wells.*

Source: Zhang, Z., Zhang, J., Du, Z., Chen, C., Zhou, Y., Sun, J., Liu, T., Zhang, J., Zhang, X., Ning, Y., & Wang, L. (2024). High-efficiency frequency doubling blue-laser VECSEL based on intracavity beam control. Sensors, 24(12), 3913. https://doi.org/10.3390/s24123913, Creative Commons License.

6.5. QUANTUM WELLS

VECSEL quantum wells (Figure 6.3) are constructed in a resonant periodic gain configuration, meaning that each well is located in an anti-node of the distribution of the electric field for the wavelength of the laser (Nakwaski and Sarzał, 2016). The standing wave periods of various wavelengths cause their field distributions to overlap with the quantum wells less precisely. There is less effective gain and a lower confinement factor. In a VECSEL, the quantity of quantum wells utilized can change.

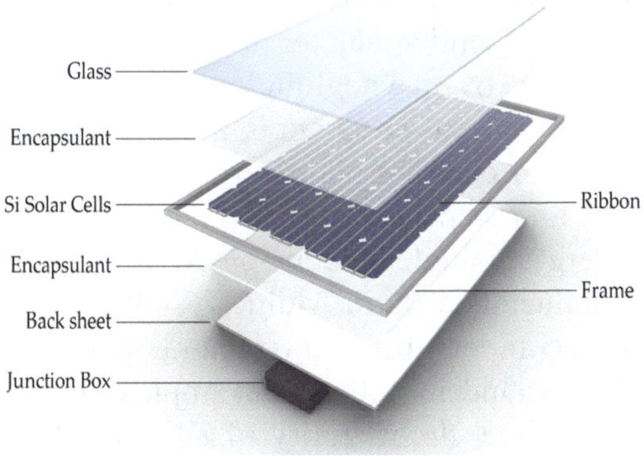

Figure 6.4. *Solar cell.*

Source: Ko, J., Kim, K., Sohn, J. W., Jang, H., Lee, H.-S., Kim, D., & Kang, Y. (2023). Review on separation processes of end-of-life silicon photovoltaic modules. Energies, 16(11), 4327. https://doi.org/10.3390/en16114327, Creative Commons License.

Higher gain is achieved with more quantum wells, but the gain structure thickens and is more susceptible to temperature effects, strain, and fabrication defects. Internal temperature gradients and varying quantum well excitation levels can lead to inhomogeneous gain saturation. In mode-locked or narrow linewidth VECSELs, this could be an issue.

The field distribution and the wavelength of the quantum well maximal intrinsic gain are both impacted by temperature variations. Gain structures are designed to obtain an optimal match of all parameters at the projected operating temperature because heating is unavoidable. Physical strain is introduced by the lattice constant mismatch between the material utilized for the Bragg mirrors and the quantum well material. Issues caused by lattice strain, such as dark lines from photoluminesence quenching, appear in the VECSEL output. Lattice strain propagating along specific crystal lattice directions complicates the issue.

Extra layers with a suitable thickness and an intermediate lattice constant help to mitigate lattice strain problems. Compressive and tensile strain are balanced by this strain compensation.

6.6. SOLAR CELL

A solar cell (Figure 6.4) is a large area np junction diode that uses the photovoltaic effect to turn sunlight into electricity. Photocurrent is produced when electron-hole pairs produced by absorbed photons are separated inside the diode. An external load circuit receives this current and transforms it into a voltage. To enable the highest amount of semiconductor surface area to be exposed to incident light, the top metal contact covers a negligible portion of the front surface area. Light absorption, electric current production, transportation of photogenerated charges, and conversion of entering photocurrent to photovoltaic power in an external circuit are the four steps that make up the conversion of incident light energy to electrical energy. A single junction diode composed of a homogeneous single crystal semiconductor is used. Multi-crystalline silicon homogeneous junction cells, thin-film compound semiconductor heterogeneous junction cells, and multi-semiconductor tandem cells are examples of solar cells (Banerjee, 2024). Near the cell surface, the majority of silicon solar cells have a high generation rate. At shorter wavelengths when α is high, the roll-off is observed, although it is exponential at all wavelengths. The photon density peaks at about 500 nm. The photocurrent is the current produced by photon absorption. The photogenerated electron-hole pairs are separated so that the electrons and holes move in opposing directions. The np junction should be positioned relatively near the front surface of the solar cell because the generation rate decreases exponentially as one moves deeper into the cell.

The semiconductor emitter is thin and the emitter region needs to be extensively doped to prevent undesired, parasitic series resistance since the current must travel in a lateral path to the metallic contacts on the front surface. The base is the lower portion of the diode. In order to collect long wavelength photons and give the solar cell mechanical strength, the base needs to be thicker than the emitter layer. Since the minority carriers produced in this area would have to travel long distances to reach the junction, the base doping type must supply the mobile minority carrier. This indicates that the base should be p type for solar cell materials like Si and GaAs. A solar cell is a np diode with a strongly doped emitter and a shallow junction. Both electrons and holes in both of the quasi-neutral zones diffuse as the production rate drops with increasing cell distance. Charge carrier separation in the emitter takes place at the junction, where the intrinsic potential barrier reflects the electrons. By strongly doping the end of the p type zone, a back-surface field is produced that produces the identical effect for carriers generated in the base. The donor density gradient, which occurs during the diffusion doping process, also creates a field at the cell front surface. The built-in junction field in the space-charge portion of the cell separates the carriers that are produced in the space charge region. The expression for the photogeneration current in this region is produced by evaluating the photocurrent due to either electrons or holes. In the space charge layer, the built-in field separates all carriers generated in that region. The photocurrent electrons and holes move in the direction of a reverse bias current. These current forward biases the diode by creating a voltage across an external load as it passes through it. There is some net current loss as a result of the development of an opposing current to the photocurrent. The photocurrent expressions are only applicable when the illumination is not so strong that the quasi-Fermi levels in the space-charge area deviate very less from constant values. The lifetime of a minority carrier remains constant. Boundary conditions at the contact edges of the quasi-neutral areas and the margins of the space charge region are used to calculate the diode dark current. The total dark current is the sum of the two injected currents, excluding any current change due to recombination in the space charge region and the hole dark current. Due to the difficulty of introducing dopants into this ternary compound crystal, vacancies are produced in the deposited film to provide extra carriers. The selenium atoms around the copper vacancies will be electron-deficient if the growth conditions are changed to create a copper deficit in the CIGS layer. The material becomes p-type when selenium atoms take up electrons from other sources. Heterogeneous junction diodes are used for the n type emitter in CIGS solar cells, which having a p type base. Since incident solar energy quickly passes through the absorbent CIGS layer, the n-type semiconductor in CIGS solar cells has a large bandgap and does not absorb much sunlight (Boukortt et al., 2020).

6.7. JUNCTION

Heterogeneous junction solar cells based on silicon are effective because a layer with a large bandgap is inserted to shift the high recombination active contacts away from the crystalline surface. The hetero-geneous / homogeneous interface state density must be as low as possible to realize the entire device potential. This is accomplished by using a hydrogenated amorphous silicon (a-Si:H) sheet that is a nanometer thick and has a large bandgap than crystalline silicon (c-Si). Through hydrogenation, amorphous silicon films lower the density of the crystalline silicon (c-Si) surface state. With very low values for the saturation current density, such films can be readily doped (n/p) type without the need for lithography fabrication of contacts. An intrinsic hydrogenated amorphous silicon (a-Si:H) passivation layer and a p-doped amorphous silicon emitter, both produced by plasma enhanced chemical vapor deposition (PECVD), are the fundamental device features on the illumination side. Physical vapor deposition (PVD) (Figure 6.5) is used to form an antireflective transparent conductive oxide (TCO) on top of the silicon layers, and a screen-printed metallic contacting grid is used to collect the charge. An intrinsic hydrogenated amorphous silicon (a-Si:H) passivation layer, a doped n type amorphous silicon, a TCO layer, and a metallic contacting layer make up the electron collecting stack that is utilized on the back side. Heterogeneous junction silicon-based solar cells can achieve efficiencies of over 25% due to low defect density in high quality (purity) silicon wafers. This high-efficiency device requires that there be no electrically active defects on the wafer surface. Plasma-deposited hydrogenated amorphous silicon (a-Si:H) produces high-quality surface passivation with long carrier lifetimes and efficiencies. Making sure the electrons and holes are collected at two terminals that are spaced apart is a major problem when creating a high-efficiency solar cell from a high-quality silicon wafer.

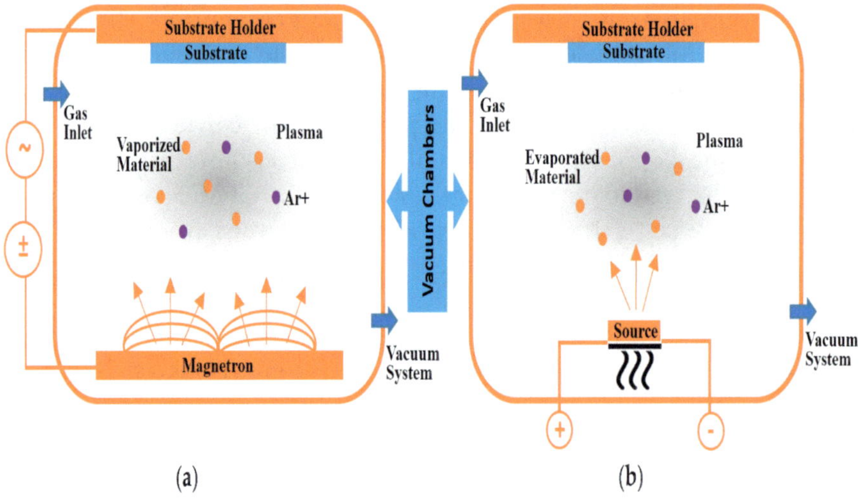

Figure 6.5. *Physical vapor deposition (a and b).*

Source: Baptista, A., Silva, F., Porteiro, J., Míguez, J., & Pinto, G. (2018). Sputtering physical vapor deposition (PVD) Coatings: A critical review on process improvement and market trend demands. Coatings, 8(11), 402. https://doi.org/10.3390/coatings8110402, Creative Commons License.

The characteristics of semi-permeable electronic membranes are exploited by selective collection. These membranes block the other type of charge (holes) with limited leakage while offering a low-resistance electrical connection for one type of charge (electrons). The best method for achieving the required charge carrier selectivity is to use doped amorphous silicon layers. But, a-Si:H films present disadvantages such as parasitic light absorption and suboptimal selectivity. The doped connections of opposite polarities are necessary for an effective photovoltaic system. Using slightly sub-stoichiometric molybdenum trioxide (MoO_3) and lithium fluoride (LiF) as hole/electron selective contacts, completely dopant-free crystalline silicon cell provides a completely new device design with simple designs and processes. A silicon solar cell electrical charges are extracted by metal contacts. The IDC design collects both positive (holes) and negative (electron) charges on the back (rear) of the wafer, whereas the conventional design of solar cells collects both charge types on either side of the wafer. Because all metal contacts are positioned near the rear side, shadowing is avoided and higher current can be produced. Wafer-based silicon devices with areas greater than 100 cm^2 have recently achieved higher efficiency (Goodrich et al., 2013). The manufacture of small devices is difficult due to edge recombination and the large diffusion length of photogenerated carriers in silicon, which is usually of the millimeter scale.

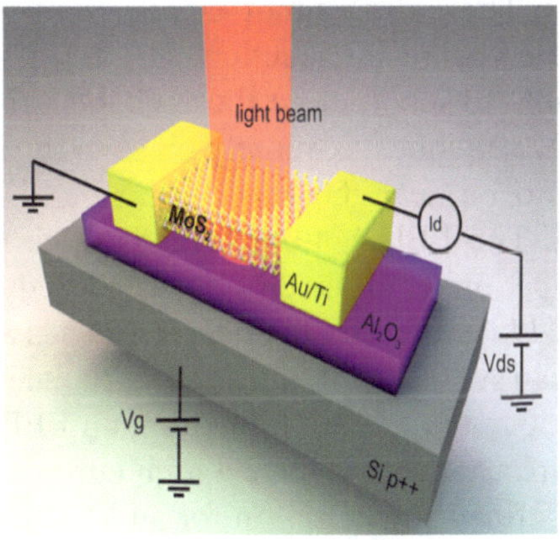

Figure 6.6. *Phototransistor.*

Source: Lavrov, S. D., & Guskov, A. A. (2025). Modeling of phototransistors based on quasi-two-dimensional transition metal dichalcogenides. Modelling, 6(2), 47. https://doi.org/10.3390/modelling6020047, Creative Commons License.

6.8. PHOTOTRANSISTOR

A phototransistor (Figure 6.6) is a photodiode extension. A light emitting diode (LED) is a p-n junction diode that, when operated under forward bias, produces light via radiative recombination. In contrast to what occurs in a solar cell, electrical energy is converted to optical energy. The top surface of most LED designs allows internally generated photons to escape since it cannot be covered by a top metallic contact. At equilibrium, holes are found near the top of the valence band while electrons are found toward the bottom of the conduction band. Although the distribution of electrons and holes is disturbed by an applied forward bias, the energy of any photons produced by radiative recombination will be around the bandgap. The local electron hole concentration product determines the rate of radiative recombination. Minority carrier concentrations in a forward biased diode are raised above their equilibrium levels over distances that are several minority carrier diffusion lengths from the space charge layer. It is necessary to concentrate the minority carriers into a smaller area in order to achieve intense photon production. The p and n-regions of the diode in high brightness LEDs are made of different materials. In order to create a potential well in which the minority carriers are trapped and raise their concentrations and the intensity of the light produced, this design employs a low bandgap material positioned between two larger bandgap materials. The active layer is the area of carrier confinement. The holes cannot escape from the active area due to a barrier at the active n region intersection. By using Type I heterogeneous junctions, LEDs are able to incorporate these characteristics. Since all of the photons in the LED have energies extremely close to the bandgap, the junction region does not need to be near the surface; hence, the relevant absorption coefficient is minimal. The LED bottom layer is thick to give it mechanical inflexibility, and n-type material is employed to reduce series resistance. Because there is limited lateral current and the LED has a significantly smaller cross-sectional area than a solar cell, vertical resistance predominates in this device. In order to achieve the required junction, an LED that runs in forward bias must avoid series resistance and have a high current. Recombination must be contained inside a defined active region for an LED to function, and it must be radiative to produce the light. A direct bandgap semiconductor should be the active material. High concentrations of both kinds of carriers are required for a high rate of radiative recombination. The LED needs to be forward biased in order to achieve a high degree of electron and hole injection into the active layer. To create the region p type, the active layer background doping density is typically selected. Since electrons are minority carriers, a more even distribution of carriers in the active layer is guaranteed by their better minority carrier mobility. Since most semiconductors have a relatively high refractive index, the critical angle beyond which total internal reflection takes place is minimal, making light extraction intrinsically challenging. GaAs has a refractive index of roughly 3.5 at the semiconductor/air contact (Thyagarajan and Ghatak, 2011). This indicates that just around 2% of the optical power would leak into the surrounding air for a point source in GaAs. The LED semiconductor layers are typically formed onto a substrate epitaxially, producing a planar structure. Due of this, a substantial portion of the light produced inside the LED may escape because the light source is planar rather than point. By effectively raising the critical angle, the semiconductor die can be sawed and converted into wedge-shaped diodes, which reduce total internal reflection.

6.9. LUMINOUS EFFICIENCY

The product of luminous efficacy and wall plug efficiency is known as luminous efficiency. White light is produced by combining the output of an InGaN LED with red light from an AlGaInP LED. Rod and cone cells in the retina of the human eye, which are particularly sensitive to red, green, or blue light, are responsible for color perception. The eye interprets combinations of these colors at specific intensities as white light.

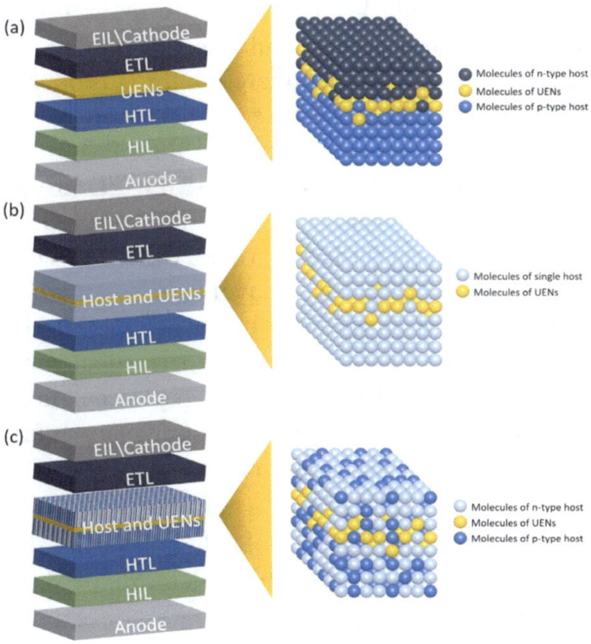

Figure 6.7. *OLED (organic light emitting diode) (a to c).*

Source: Zhou, Y., Gao, H., Wang, J., Yeung, F. S. Y., Lin, S., Li, X., Liao, S., Luo, D., Kwok, H. S., & Liu, B. (2023). Organic light-emitting diodes with ultrathin emitting nanolayers. Electronics, 12(14), 3164. https://doi.org/10.3390/electronics12143164, Creative Commons License.

An organic light emitting diode (OLED) light-generating mechanism is simple (Figure 6.7). The cathode and anode both receive electrical current. Current flows to the anode layer via the conductive and emissive cathode and anode layers of organic material. Electrons are transferred from the conductive layer to the emissive layer by the current. The conductive layer develops holes. Electrons must be added to the conductive layer to fill the holes that have been created. The holes move from the conductive layer to the emissive layer, which is filled with electrons, in order to recombine with them. The visible portion of the electromagnetic spectrum, electroluminescent light visible through the outermost layer of glass, is produced by electron hole recombination. Because of its organic nature, the OLED self-illuminates rather than using a backlight like LEDs do. OLEDs are thinner than conventional liquid crystal displays. There are three types of OLED displays: transparent OLEDs, active mode (AMOLED), and passive mode (PMOLED). The passive mode (PMOLED) has a simple physical design. The emissive and conductive layers are positioned between the anode and cathode layers, which are made up of narrow strips that are a nanometer wide and perpendicular to one another. At the locations where the cathode and anode layers converge, light is produced. To create an image, the TFT (Thin film transistor layer) array controls which pixels are turned on. AMOLED are more energy efficient than PMOLED because TFT arrays use less power than external circuitry. AMOLED are perfect for large displays because they have fast refresh rates than PMOLED. Military aircraft head up displays use transparent OLED (Sarma, 2015). The OLED is over 85% transparent when turned off. Light may travel both ways when it is powered on. Transparent OLED can be either an active or passive matrix. The pilot of a military aircraft can examine important aircraft performance parameters while looking directly ahead without having to shift his or her head up, down, or sideways. An extension of a photodiode, a phototransistor equivalent electrical circuit is made up of a photodiode whose output photocurrent powers a small signal transistor base.

From near UV wavelengths through the visible and near infrared regions of the spectrum, the wavelength of input light determines the phototransistor output. At about 840 nm, the highest spectral response occurs in the near infrared without optical filters. Since the diffused junctions of a phototransistor are generated in epitaxial silicon wafers rather than crystal-grown, the peak response is at a shorter wavelength than that of a typical photodiode. Infrared LEDs such as GaAs (940 nm) and GaAlAs (880 nm) exhibit superior optical coupling efficiency when paired with phototransistors, which will react to fluorescent or incandescent light sources (Gupta and Gupta, 2016). The area of the exposed collector base junction and the transistor DC current gain determine a phototransistor output for a specific light source illumination level.

The phototransistor collector base junction produces a photocurrent that is fed into the transistor base, acting as a photodiode. Doubling the base size produces twice as much base photocurrent (IP), which is subsequently amplified by the transistor DC current gain, much like in a conventional photodiode. Similar to signal transistors, bias voltage, and temperature affect hFE or forward DC current gain. The gain begins at low light levels and rises as light levels rise until it reaches a peak. The phototransistor gain begins to drop as the light level is raised further. As the collector-emitter voltage rises, so does the DC current gain. The collector current (IC) of a phototransistor is linear for only 3–4 decades of illumination, in contrast to a photodiode whose output is linear with respect to incident light over 7–9 decades of light intensity. This is because the DC gain (hFE) of the phototransistor is a function of the collector current (IC), which is in turn determined by the base drive. The base drive could be an incident light base drive current. The phototransistor functions as a switch, but photodiodes work when linear output vs light intensity is crucial, like in light intensity measurement devices. A phototransistor or photodarlington can be said to be ON when there is light present, meaning that it can drain a significant amount of current. These photodetectors go into an OFF state and operate electrically like open switches when the light is turned off. A specific amount of photocurrent, known as the dark current (ID), will flow when the phototransistor is exposed to dark and a voltage is connected from collector to emitter (Guo et al., 2017). This current is the result of multiplying the transistor DC current gain by the leakage current of the collector-base junction.

6.10. PHOTOTRANSISTOR

The phototransistor is never fully OFF, making it an inappropriate switch for close-open transitions. The highest collector current allowed to flow at a specific collector emitter test voltage is known as the dark current. The applied collector emitter voltage and ambient temperature both affect the dark current. A phototransistor response speed is entirely determined by the load resistance and the capacitance of the collector base junction. The Miller Effect, which doubles the RC time constant by the phototransistor current gain, is the reason these predominate. For devices with the same active area, the photodetector speed of response will decrease with increasing gain. It takes time for a phototransistor to react to sudden variations in light intensity. Risetime is equal to falltime as long as the light source that powers the phototransistor is not strong enough to induce optical saturation, which is defined by the accumulation of too many charge carriers in the base region. tF (risetime) can grow significantly larger than tR (falltime) if optical saturation takes place. For phototransistors to function, they need to be appropriately biased. In particular, applied voltages must not be higher than the emitter-collector breakdown voltage (V_{BRECO}) or the collector emitter breakdown voltage (V_{BRCEO}) in order to prevent irreversible damage to the phototransistor. Silicon and germanium were initially used to make phototransistors; gallium arsenide is now a common material (Gholipour et al., 2024). Photo field effect transistors are also available. Due to their narrow operating wavelength ranges, lower quantum efficiency and detection bandwidth, phototransistors are not as commonly utilized as photodiodes. Despite having a higher responsivity than a photodiode, the phototransistor amplifies both photocurrent and dark current, hence this does not convert into lower noise equivalent power. The detection bandwidth is decreased by the large base collector, resulting in rise-fall times that are in microsecond range. Avalanche breakdown is the mechanism by which avalanche phototransistors generate current, while having a higher responsivity than regular phototransistors. Voltage-controlled devices are known as field effect phototransistors, or photoFETs. The transistor is turned on by a voltage created by light striking the gate region.

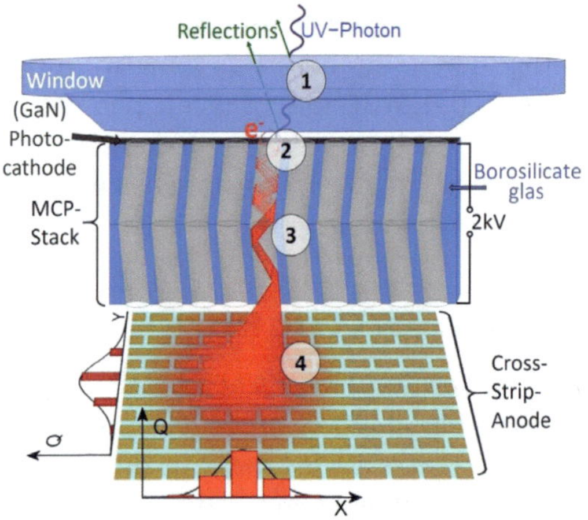

Figure 6.8. *PhotoFET.*

Source: Rogalski, A., Bielecki, Z., Mikołajczyk, J., & Wojtas, J. (2023). Ultraviolet photodetectors: From photocathodes to low-dimensional solids. Sensors, 23(9), 4452. https://doi.org/10.3390/s23094452, Creative Commons License.

A photoFET (Figure 6.8) can also be triggered by infrared radiation, therefore the incident electromagnetic radiation need not be in the visible wavelength range. By integrating incident photons over extended exposure times, an optical device, such as a telescope, allows the user to see distant objects. The speed at which photons enter the retina affects how the human eye reacts.

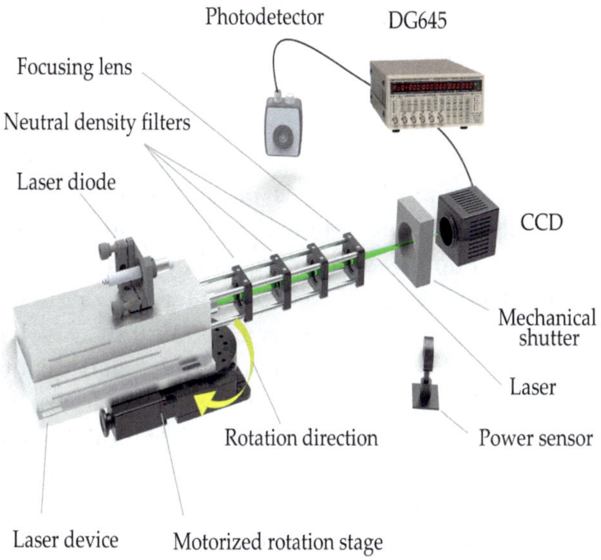

Figure 6.9. *Charge coupled device.*

Source: Yu, C., Ye, J., Chang, H., Li, N., & Guo, W. (2024). Temporal and spatial distribution characteristics of crosstalk lines generated by irradiating progressive scan charge-coupled device camera with continuous laser. Sensors, 24(12), 3966. https://doi.org/10.3390/s24123966, Creative Commons License.

Electrons are ejected from the charge coupled device (CCD), (Figure 6.9) which is the surface on which photons are incident, by the photoelectric effect at optical frequencies and wavelengths. The ratio of detected photons to incident photons is known as a light detector quantum efficiency. Throughout the whole optical frequency range, CCDs have quantum efficiencies of 10 to 90%, which can be increased with the help of efficient methods. A CCD is made up of a thick layer of p type silicon on top of a thin layer of n type silicon. Fixed positive charges are produced in the n-doped zone and fixed negative charges are found in the p-doped region. A CCD n type substrate is devoid of electrons prior to radiation exposure. An undepleted region forms at the bottom of the p-type substrate as the positively charged free charges in the p-type silicon are repelled by the positively charged n-type substrate on top of it. Because of free holes, conduction is possible. However, the absence of free charges stops conduction above this undepleted area and in both the n type and p type silicon layers. The high positive charge in the n type region, which has a higher concentration of fixed positive charges than in the p type region, causes an electric field to move downward in the p depletion layer. The devices are now prepared to catch the photoelectrons that are driven upward into the higher n layer, where they are trapped by the positive charge.

6.11. CCD

For conventional CCDs, one electron is produced for every photon that is detected. Incident photons produce photoelectrons. With a band gap energy of 1.14 eV, silicon absorbs photons with energies between 1.1 and 1.4 eV, which gives valence electrons the energy they need to enter the conduction band (Ji and Abshire, 2004). The path length of light as a function of wavelength in silicon can be used to calculate the wavelength sensitivity and quantum efficiency of CCDs. The width of the CCD also affects the quantum efficiency. The backside illuminated CCDs, which are only about 15μ thick, and the front side illuminated CCDs, which are about 300μ wide, absorb photons of incident light with wavelengths >0.35μ and <0.8μ. 70% of the incident light is reflected at wavelengths shorter than 0.35μ. The sensitivity can be extended to 0.3 μm with an anti-reflection coating, although the CCD is visible at wavelengths greater than 1 μ. In contrast to their usual behavior, the photoelectrons discharged into the conduction band do not recombine with the silicon. The electric field rapidly pushes these photoelectrons into the n type layer, where they become trapped in the potential well. The integrated light intensity at the end of the observation determines how many electrons were stored. A pixel's well capacity is the amount of charge it can retain following integration. Integration timings need to be adjusted to avoid going over the detector well capacity. Potential is applied across the top n type layer to create picture elements, or pixels. In order for the electrons to assemble at the locations within the array where the potential is highest, each pixel typically has three electrodes with varying positive voltages. The applied voltage between the electrodes can be raised or lowered between a higher and lower threshold value. The potentials are sequenced to transfer the charges along a parallelly connected column in the direction of an output register which is an unexposed row of pixels at the end of the exposed chip, in order to read out the CCD. By simultaneously changing the voltages in the three gates in every pixel, the electrons are shifted to the output register. The electrons go to the neighboring pixel in different voltage adjustment steps. The charges on the output register row pixels are shifted, likewise with changing voltages, in the direction of the output electronics when a whole row has been moved into this output register row. The well capacity, the digitization rate at which raw pixel data is transformed into usable data, and the electron capacity, beyond which the CCD output becomes non-linear (noise), affect how many photons may be detected (Pawley, 2006). The beginning of readout noise is quite hard to detect. In order to avoid amplifier heating (increased current), the readout duration is usually limited to 50 microseconds per pixel. This reduces readout noise by about 10 electrons per pixel/readout. In addition to noise currents unique to semiconductor devices, such as thermal noise, shot noise, etc., readout noise also occurs during the analog to digital conversion process. By binning the pixels before readout, readout noise can be reduced. The signal is then impacted by a single readout noise instead of four times the value if four pixels are binned. Even with high signals, binning is feasible since the output register non-illuminated pixels are typically 5 to 10 times as many as the illuminated pixels.

Because analog to digital converters cannot handle negative integers, a DC bias is included when converting CCD images from analog to digital to ensure that the signal is never negative. Instead of using the minus sign, the signal is kept positive so that the A/D converter can use all of its bits. Position across the CCD and readout time both affect the bias value. Overscan strips or a bias frame measurement can be used to determine the bias. After the full CCD image has been read out, overscan strips are put to the CCD as columns that are recorded and read out without any signal. They are deducted from each image pixel in that row after being averaged for each row, providing a measurement of the zero signal point. Recording bias frames, that is, photos shot with the shutter closed and no integration time, is another method of removing bias. The zero point in the data is represented by these pictures. Delta sigma conversion is a quick and effective analog to digital conversion method. This approach sacrifices temporal accuracy for amplitude; if the amplitude of the input signal crosses from positive to negative, the output signal amplitude is zero. The output signal amplitude is zero when the input signal amplitude is negative. Since each excited atom in spontaneous emission is free to return to the ground state anytime, the radiation emission is out of phase. The relaxed atom releases the excess energy as radiation once it reaches the ground state. All of the excited atoms in stimulated emission return to the ground state simultaneously, releasing the additional energy simultaneously, in phase. This is light emitted by a laser, which is a stimulated emission of radiation. A laser is a device where the populations of the initial and final states are raised to values that deviate from the values of thermal equilibrium using external techniques. The excited state atoms in a typical laser are permitted to return to a state with an energy value that lies in the middle of the excited state energy and the final ground state energy. The main characteristic of the intermediate state is that its atoms have extended mean lifetimes, such as milliseconds. Many atoms accumulate in this intermediate (metastable) state before simultaneously returning to the ground state, a phenomenon known as phase emitting stimulated emission.

6.12. Crystallographic Phases

There are two fundamental crystallographic phases of gallium nitride: the face-centered cubic (FCC–Zinc Blende structure) and the hexagonal close packed (HCP) (Mishra and Singh, 2008c). For each crystal, like GaN, the unit cell is made up of an arrangement of constituent atoms in three dimensions, such as nitrogen and gallium, which, when translated in the right directions, may recreate the entire crystal. Hexagonal close packed (HCP) and face centered cubic (FCC) crystal formations have distinct characteristics. Although FCC GaN is perfect for some applications, it is not suitable for others, where HCP is better. A metastable, β-phase form of gallium nitride, the FCC crystalline structure can be stabilized in epitaxial layers. The (111) closely packed planes in this structure have the stacking sequence ABCABC (Sato et al., 1967). These hexagonal and cubic phases can coexist in epitaxial layers. Inversion symmetry is absent in both GaN alpha and betaphases. The FCC crystalline structured GaN is also utilized in specific applications, even though the HCP lattice structure GaN is the preferred material for the semiconductor industry since it can be formed on sapphire substrates without lattice constant mismatch. Compared to HCP lattice structure GaN, FCC lattice structure GaN has a slightly smaller bandgap and a higher saturation electron drift velocity.

CHAPTER 7

PACKING OF SEMICONDUCTOR LASERS

Contents

7.1. INTRODUCTION

The packaging structures of semiconductor lasers are separated into two groups: fiber-coupled modules and open packages. Four components make up the fundamental design of a semiconductor laser: a mounting substrate, a bonding solder, a heat sink, and a laser chip. A laser chip may have a single emitter or several, like a bar. The laser chip is bonded to the mounting substrate using bonding solder, which is located beneath the chip. There are two types of bonding solder: soft and hard. The laser chip is mounted on the mounting substrate underneath it. Dissipating heat from laser chips and acting as a mechanical foundation are the uses of the heat sink, which is attached beneath the mounting substrate. Because of its purpose, the heat sink is sometimes referred to as a cooler. In some cases, the heat sink and mounting substrate are made of the same material. Conduction cooling and liquid cooling are the two cooling methods used for semiconductor lasers. A single emitter output power is typically low, and its heat emission is also moderate. The majority of single emitter lasers use conduct cooling.

7.2. SINGLE EMITTER SEMICONDUCTOR LASERS

Single emitter semiconductor lasers are primarily packaged in four different ways: C-mount, B-mount, CT-mount, and F-mount (Liu et al., 2015b). The cathode, a laser chip, a ceramic slice, and a mounting substrate that serves as the anode are the four major components of a C-mount. The C-mount can be mounted to an installation platform using the hole in the mounting substrate. The cathode and anode are separated by the ceramic slice. The cathode fly-lead is wire-bonded to the laser chip n-side. A C-mount is mounted on an installation platform at the side of the mounting substrate. In order to reach the platform, the heat conducts vertically and then horizontally. This indicates that when it functions, heat is first transferred vertically to the mounting substrate and subsequently horizontally to the platform for installation. Beneath the chip, the two pathways that run in vertical and horizontal directions overlap. The overlapped path makes it impossible for heat to be dispersed effectively, necessitating the optimization of the heat conduction channel. The B-mount is composed of four components: the cathode, a laser chip, a ceramic slice, and a mounting substrate. Similar to the C-mount, the B-mount mounting substrate serves as both the anode and a heat conductor in the single emitter laser. The cathode fly-lead is positioned on a ceramic slice. The fly-lead is shielded from the mounting substrate by the ceramic slice. The laser chip n-side is wire-bonded to the cathode. The direction of heat conduction is vertical. The distinction between these two lasers and the C-mount laser is that the CT-mount laser allows for a heat sink to be placed immediately beneath the mounting substrate. Like the C-mount, the CT-mount cathode is wire-bonded to the laser chip n-side, while the anode serves as the mounting substrate. The laser chip, the ceramic mounting substrate, the cathode, the anode, the ceramic slices, and the heat sink are the six components that make up the F-mount. The anode and cathode are separated from the heat sink using the ceramic slices. The heat sink in the F-mount is insulated from the laser diode and is not charged when the laser is operating. The ceramic mounting substrate has a gold layer finish applied to its surface. The insulation grove separates the ceramic mounting substrate into two sections, denoted by the letters + and – respectively, for the anode and cathode. The bonding solder layer adheres the laser chip to the ceramic mounting substrate. Wire bonding is used to join the cathode to the laser chip n-side. The two fly-leads, which are the anode and cathode of the F-mount, are wire-bonded to the anode and cathode of the ceramic mounting substrate. The ceramic mounting substrate has a thermistor pad that can be used to regulate the laser diode temperature. Even in CW operation, the copper conduction cooler can readily dissipate several watts of heat generated by a single emitter semiconductor laser. However, while diode laser bars are operating, tens or even hundreds of watts of heat will be produced. For diode laser bars, the cooler structure needs to be tuned; in certain situations, like the heavy-duty cycle and CW mode, liquid coolers are required. The micro-channel cooler (MCC) and the macro-channel cooler (MaCC) are the two varieties of liquid coolers. According to the liquid cooler, four essential components can be used to build bar-based semiconductor lasers: an insulator, a cathode plate, a laser bar, and a mounting substrate or cooler. A Cu/CuW heat sink, an MCC, or a MaCC can be the mounting substrate or cooler. Conduction-cooled semiconductor laser bar (CS) refers to a laser bar that has a conduction heat sink, such as Cu or CuW (Jiang et al., 2013).

7.3. SINGLE-BAR SEMICONDUCTOR LASER

In Single-bar semiconductor lasers cooled by conduction, along with CS, the conduction-cooled single bar package is also referred to as CCP in the production process. The solder layer between the heat sink and laser bar is extremely thin and difficult to observe. The laser chip is epi-down mounted on the heat sink. The heat sink serves two purposes and is a large and thick copper block. It serves as the laser bar anode in addition to being a cooler for heat dissipation. The device is installed on a platform using the mounting holes, which are the four large holes in the heat sink. The anode thread is located in the small hole between the large holes on the left side of the heat sink, and a thermistor is fixed in the small hole on the right. The CS cooler is significantly larger than the laser bar in order to achieve adequate heat dissipation. A smaller cooler size known as NCCP was developed for conduction cooling laser devices with narrow or short duty cycles. Single semiconductor laser bars that are liquid-cooled have cathode, anode, FAC, and laser bar. The MCC and MaCC are typically several millimeters height and 10 mm wide, though they can be the same size. It serves as the laser bar anode in addition to being a cooler. The cooler channel designs are the primary structural difference between the MCC and MaCC lasers. The high power semiconductor laser bar with MCC can function under CW and

QCW with a high duty cycle, and the micro-channel cooler has a high heat dissipation efficiency. The MCC coolant inlet is situated close to the laser bar. There is a small hole between the coolant input and outflow that is used for installation. Three layers make up the MCC laser. The cathode is the topmost layer. The microchannel, which serves as the laser bar anode, is the down layer. The insulator layer, which divides the cathode from the anode, is located in the middle layer between the cathode and micro-channel cooler. A thin layer of bonding solder is used to package the laser chip on the micro-channel cooler. The structure of MCC itself is complicated, consisting of five copper layers. The micro-channel layers, which make up the second and fourth layers, have a width of several micrometers. Micro-channel cutouts in extremely thin copper foils are used to produce MCC. If the coolant liquid contains excess debris, the micro channel can get blocked. If there is an excess of electrolyte in the liquid, the channels are vulnerable electrochemical corrosion. Only deionized water should be used; tap or industrial water cannot be utilized. A cathode, a MaCC cooler, a laser chip, a solder layer, and an insulator make up a MaCC semiconductor laser. A standard MCC cooler is composed of five thin copper foils. However, the MaCC cooler is made from a single, 1.5–1.6 mm thick sheet of copper. The coolant inflow is separated into small channels, each of which is significantly larger than the MCC channel. Deionized water is not required; industrial tap water can be used instead. Compared to MCC, the cooling cost of MaCC is significantly lower. MaCC is simpler to fabricate than MCC because of its simple structure. However, high power MaCC semiconductor lasers operate under QCW with low duty cycle because MaCC heat dissipation efficiency is significantly lower than MCC, particularly at high power (Wang et al., 2010).

7.4. PACKAGE STRUCTURES

Semiconductor laser bar array package structures have been designed to achieve high power. The horizontal array, often known as the H-array, is a laser array with bars oriented horizontally. The laser array known as the vertical stack (V-stack) has the bars are stacked vertically. Liquid coolers are typically used in commercial H-array and V-stack semiconductor lasers. Conduction-cooled semiconductor laser stack (G-stack) is the semiconductor laser bar array that uses a conduction cooler. While the G-stack semiconductor laser can only operate at QCW with a low duty cycle, the H-array and V-stack semiconductor lasers can run at both CW and QCW. The H-array packaging structure is the configuration for H-array lasers with four and three laser bars. The output beam is a line beam since the laser bars are oriented horizontally. As the number of semiconductor laser bars increases, the beam product parameter (BPP) of H-array lasers worsens. H-array lasers are favored in certain applications where a line beam is required, such as the side pumping of solid state lasers, and there is no theoretical limit on the number of laser bars in the array. Although the H-array lasers have distinct packaging configurations, they are made up of semiconductor laser bars, a cathode, an anode, a heat sink, and coolant input and outlet. The cooler's coolant outlet is located next to the cathode and its coolant input is close to the laser array anode. H-arrays are not recommended for applications where high beam quality is necessary because of the declining beam quality along the slow axis. This drawback of the H-array laser is addressed by the vertical stack semiconductor laser. V-stack lasers maintain the same beam quality along the slow axis as a single laser bar. A V-stack laser maximum output power can range from several 100 watts to up to 20,000 watts, and the number of laser bars in a V-stack laser has expanded to 70 bars due to advancements in packaging technology. Either MCC or MaCC laser bars are available for the V-stack semiconductor laser (Wang et al., 2015). The key component includes the packaged MaCC (or MCC) bars, the cathode, the anode, and the coolant input and outlet. The MCC V-stack cathode and anode has the coolant outlet and intake. The coolant outlet and inlet are both located at the anode of the MaCC V-stack laser. Similar to MCC and MaCC lasers, MCC V-stack lasers can operate in both CW and QCW modes, but MaCC V-stack semiconductor lasers can only operate in low duty cycle QCW mode. In G-stack packaging structure, the insulator layer sits between the conduction cooler and the laser chips, while the heat conduction cooler is beneath the laser chips. The cathode and cooler and the anode and cooler are separated by two insulator layers, respectively. The stack has four fixing holes, which serve to secure the stack and link the stack laser anode and cathode to the power driver, respectively. Many alternative G-stack laser structures have been designed to simplify the package structure. The advantages of the G-stack laser over the H-array and V-stack lasers are its high peak power, small volume, and simplicity of integration. However, they can only operate in low duty cycle and QCW mode. The packaging structures of the fiber-coupled modules are intricate. They have a module house, cooler, fiber ferrule assembly, optical beam shaping, and fiber-coupled optics in addition to the laser chips or laser bars. A feedback mechanism is also required in certain applications. The module comprises thermoelectric cooler (TEC), the optical system (OS), module body, fiber ferrule assembly, and solder in addition to the single emitter, which is the F-mount. The beam of a single emitter module can be linked into 200 μm fiber due to its small beam size. The feedback systems of a fiber-coupled module have many varieties. Temperature sensors, fiber detection, power photoelectric detectors (PD), red indicators are examples of feedback systems. Single emitter semiconductor lasers have a simple package structure. The fiber module can be created for a particular application even without TEC and feedback mechanisms.

7.5. MODULES

The beam shaping optical system, pigtail, multiple single emitters, and module body make up the structure of multi-emitter fiber-coupled modules. The module beam shaping optical system couple laser beams into optical fiber and enhances beam quality. The module has three emitters that are separated spatially by the steps, and the pigtail and module body have a simple structure. Three mirrors, two cylindrical lenses to concentrate the laser beam, and three collimator lenses along the slow axis make up the beam shaping optical system. This module typically has a conduction cooling packed semiconductor laser bar with emitters.

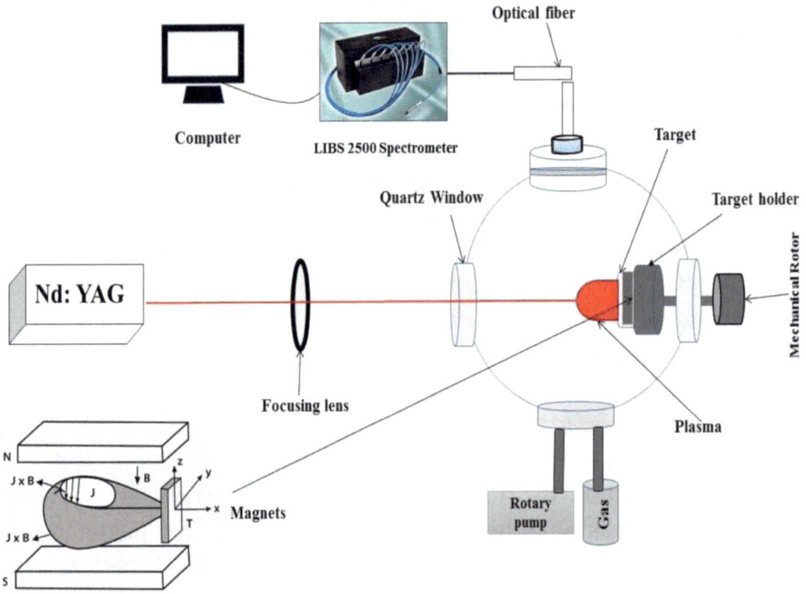

Figure 7.1. *Conduction-cooled semiconductor lasers.*

Source: Dawood, A., Ahmed, N., Bashir, S., Hayat, A., Sarfraz, S. M. A., & Ayub, A. (2022). Laser ablation of copper alloy under varying environmental conditions to achieve purpose-built surface structures. Coatings, 12(12), 1972. https://doi.org/10.3390/coatings12121972, Creative Commons License.

The laser beams are connected into an array of fibers following fast-axis collimation for the emitters. The superposition of the beams linked in the fiber equals the output power. Accurate alignment between the semiconductor laser and array fiber is a challenge in the fabrication of the module. The beam shaping system in the module body is the primary distinction between the two types. Only beam collimation along the fast axis is required prior to fiber coupling in the fiber bundle mode, and the beam does not require complicated beam shaping. However, it requires a sophisticated beam shaping technology in addition to beam collimation; this type of device is known as a free space fiber-coupled structure. Compared to the fiber bundle module, the module is brighter.

7.6. CS

Three conduction-cooled semiconductor lasers (CS) (Figure 7.1) are installed on the ladder platform and are part of the module. Each CS device has a piece of ceramic placed underneath it for insulation (Ma and Zhong, 2008). The beam must be collimated and the divergence angles decreased by adding a fast-axis and slow-axis collimation lens. The two main methods used to improve the beam quality in the multiple bar-coupled module are optical coupling and beam shaping. The most common beam treatment techniques are wavelength combination, polarization combination, and spatial combination, which are generally considered to be beam treatment methods. Higher fiber coupling output power can be achieved using these techniques. The beam is shaped into the fiber using two cylindrical lenses. Since the junction temperature increase caused by heat fluxes has a significant impact on the device wavelength, power, efficiency, and reliability, thermal management of high-power lasers is essential. A few thermal management strategies are suggested to lower the diode laser device junction temperature. These strategies include cooler design, cooling structure optimization, and packaging material optimization. Compared to other laser types, high power diode lasers have a substantially better conversion efficiency. The conversion efficiency is above 55% for 808 nm and 65% for 980 nm, whereas the laboratory conversion efficiency is close to 75% to 80% (Wang et al., 2017). However, high power diode lasers still produce a substantial quantity of heat because of their high output power, which raises the temperature at the diode laser active region. Lowering the heat sink temperature is the initial strategy. Reducing the laser device thermal resistance is the second strategy. Increasing the optical output power is the third method. Raising the temperature in the diode laser active zone lowers carrier confinement and raises the likelihood of non-radiative recombination. A higher threshold current and a decreased conversion efficiency are the results of both impacts. The active region temperature increase limits the maximum output power. The high temperature of the diode laser active zone has a detrimental effect on other crucial factors like wavelength, spectrum, and lifetime. The barrier height and the required threshold-current density, which are based on the resonator losses and the optical confinement factor for a particular active region, determine the characteristic temperature T_0. At 980 nm, the values for T_0 are approximately 200 K; at 800 nm, they drop to about 140 K; and at shorter wavelengths (630–

760 nm), they fall to 100 K. A typical single emitter diode laser threshold varies with operating temperature. The threshold rises as the temperature rises.

7.7. SLOPE EFFICIENCY

Slope efficiency is influenced by internal quantum efficiency, front and rear facet reflectivity, absorption in the semiconductor material, and photon emission frequency (Dowell and Schmerge, 2009). The slope efficiency varies with operation temperature. The output wavelength red shift as the emitter junction temperature rises. A semiconductor laser wavelength will change with junction temperature at a specific rate. The lifetime testing data can be used to examine the semiconductor laser lifetime. Lowering the junction temperature is the simplest and efficient method to extend the life and boost device dependability. Heat may be produced at the materials or interfaces of the packaging when the diode laser chip is assembled into a device. A semiconductor laser device has various heat sources. They are in the active zone (non-radiative recombination), radiation absorption within an optical cavity, radiation absorption beyond an optical cavity, joule heating throughout a laser structure layer, and joule heating on the surface at semiconductor metal interface. The diode laser chip is responsible for the Joule heating at semiconductor metal interface. The structure and design of the laser chip determine how much heat is produced by these heat sources, and the computation is intricate. The laser device package structure has a significant impact on its characteristics. In C-mount, the bonding wires, solder, chip, and mounting substrate are the primary components of a C-mount packaged device. Q1, Q2, Q3, and Q4 stand for the heat produced by these components, respectively. In addition to heat produced by these bulk materials, heat can also be produced at the interfaces between these components. There would be voids at the interfaces and a high electrical contact resistance if the solder between the adjacent layers was not well bonded, which would cause the laser device to operate with excess heat. The improper contact between two surfaces produces the surface Joule heating Q5 and Q6. Choosing the accurate materials and refining the soldering technique can strengthen and enhance the contact between two surfaces. Surface resistance would be extremely low and surface Joule heating could be disregarded if the solder interface and soldering procedure were adjusted. In semiconductor lasers, metallization and an improved soldering technique are used to make contact between the solder layer and the Cu heat sink and between the solder layer and the laser chip. Since the surface Joule heating is often relatively low, Q5 and Q6 can be ignored in this instance. Q2 heat source is made up of multiple heat producing methods. When creating a high-power semiconductor laser package, thermal modeling and design are crucial stages.

7.8. MODELING

When performing thermal modeling and design, we need to maximize package structure from the perspective of thermal management, determine the resistance to heat, determine the active region maximum temperature, and examine heat flow and temperature field distribution to induce both steady state and transient thermal behavior. Finite element modeling (FEM) has become a very useful technique for thermal design and analysis of semiconductor laser packages as computer technology has advanced. The fundamental idea behind FEM is that the laser device is broken up into many small pieces, and the thermal characteristics of each unit are computed. The overall thermal performance of the laser is attained by combining the outcomes on each unit. A number of commercial software programs have been created to simulate a device thermal characteristics, including ESC TMG, COSMOS, Ansys, and ICEPAK (Monier-Vinard et al., 2017). One of the most often used programs in this industry is Ansys. The precise construction of a semiconductor laser device, which includes building each layer of the structure, is required in the solid model step. Each layer dimensions, composition, and arrangement must match those of the actual laser device. It is important to note that the chip layer structure model should, to the extent feasible, match the laser chip actual structure. The CS device structure is separated into multiple sections. The cathode, laser chip, solder layers, and mounting substrate are the primary components. The laser chip is the most important component that allows the laser device to precisely stimulate its heat.

Following the construction of the solid model of the semiconductor laser, the solid model must be divided into several small grids. We refer to this process as meshing. Rougher grids can be used at less crucial sites, like the mounting substrate and electrode, while finer grids must be built to precisely depict the fine structure at the crucial areas, like the solder layer and laser bar, during the meshing process. While the other components of the solid model are meshed with rougher grids, the laser bar, which includes the solder layer, cladding layers, metallization layers, and quantum well layer, is meshed with finer grids. The diode laser heat generation and the laser device ambient temperature are examples of boundary conditions. In order to create a successful simulation that is consistent with the device actual temperature distribution, boundary conditions are crucial throughout the FEM analysis. The original temperature distribution and the stimulated object heat flow density are examples of boundary conditions. The p-side connection is made by epi-down mounting the laser chip onto a Cu or CuW substrate. The n-side connection is made to the substrate by a fly-lead. The fly-lead and the chip n-side are wire-bonded together. When the laser is turned on, the heat is first transferred vertically to the heat sink and then horizontally to the cooler. This deviated thermal path might raise the device junction temperature and further affect its performance because it is ineffective at dispersing heat produced by the chip. Another package type for semiconductor lasers with a single emitter is the F-mount. The ceramic mounting substrate (AlN) is used to mount the flip-chip. The substrate that is placed on top of the cooler has the n-side wire-bonded to it. The heat produced by the chip can be readily transferred through the mounting substrate to the heat sink when the laser is turned on.

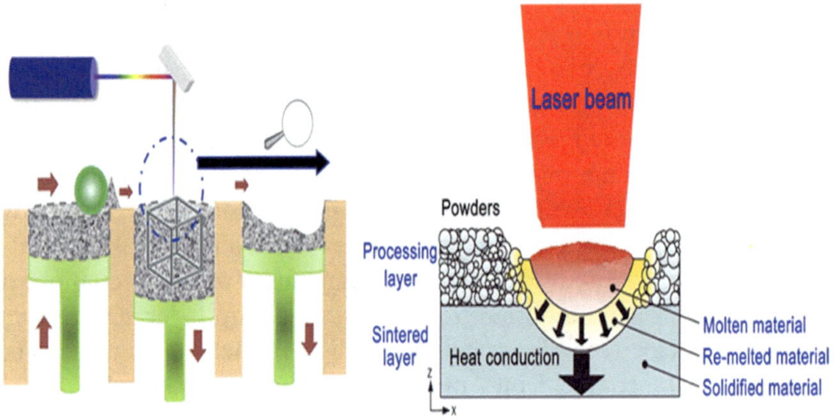

Figure 7.2. *F-mount laser package.*

Source: Yan, A., Atif, A. M., Wang, X., Lan, T., & Wang, Z. (2022). The microstructure and cracking behaviors of pure molybdenum fabricated by selective laser melting. Materials, 15(18), 6230. https://doi.org/10.3390/ma15186230, Creative Commons License.

7.9. F-MOUNT

The F-mount laser package (Figure 7.2) can increase thermal dissipation efficiency (Li et al., 2012). The bandgap of the laser active region, which is affected by temperature and thermal stress, is known to directly determine the wavelength. The wavelength difference is unaffected by heat accumulation when lasers are run in QCW mode with a low duty cycle and small pulse width because almost no heat is produced on the laser chip during operation. However, when lasers are run in CW mode, a lot of heat is produced, and the thermal effect causes the wavelengths of both kinds of lasers to redshift. This is because the C-mount heat dissipation efficiency is lower than the F-mount, causing more redshift in the C-mount. The active zone, which contains emitters is known as the quantum well or QW. This laser bar serves as the foundation for the modeling and analysis. FEM was used to simulate the stable thermal behavior of a semiconductor laser in CW mode that was packaged in a CS. The device generates heat through the emitters in the quantum well. When the device is operating in continuous wave mode, its output power is 60 W. The device rear side is maintained at 25°C. The maximum temperature in the active region is 51.8°C when the device is operating in continuous wave mode with an output power of 60 W. The transverse temperature profile of quantum wells in the steady state is essentially distributed in the form of a parabola. The temperature increases per unit length in emitter zones are higher than in other parts because heat movement is restricted in a small region. As time progressed, the heat moved from the active zone to the heat sink. The boundary condition is identical to the one found in the thermal analysis of stable states. Heat spreads to nearby emitters and thermal crosstalk occurs between emitters during the 200–300 μs time. When thermal crosstalk occurs, each emitter temperature distribution is impacted by some factors. The heat produced by the emitter itself makes up one portion, and the overlap of heat transferred from neighboring emitters makes up the other factor. The heat produced inside the active zone travels to the upper side of the solder layer in roughly 280 ns. The transmission time to the rear side of the solder layer is roughly 900 ns. When the pulse width is less than 280 ns, the solder voids would not affect the temperature distribution in the quantum well in any way. Solder voids would progressively alter the temperature distribution in the quantum well for pulses longer than 280 ns (Liang et al., 2007). Solder voids would have an impact on the laser device thermal performance and the junction temperature distribution for pulses longer than 900 ns.

7.10. CW OPERATION

A typical laser dependability is only good under regulated conditions, such as during short pulse QCW or CW operation (Stoltz et al., 1998). Its dependability quickly declines when it operates on long pulse QCW operation. This can be explained by the thermal simulation. When a semiconductor laser operates in the relatively short pulse QCW mode, the solder layer is not affected by temperature changes and does not have enough time for heat to transfer. The temperature cycling does not affect the solder layer. Heat is transferred past the upper side of the solder layer and inside the solder layer when the device operates in the larger pulse width, but not to the lower side. There would be a significant temperature difference in the solder layer, which could produce significant electromigration, particularly when the driving current is high. The solder layer, however, hardly experience the pulses or the temperature change they cause for this range of pulse width. Solder thermal fatigue is not a major concern because of the mild temperature cycling on the solder layer. The solder layer temperature distribution is already constant within a single pulse when the device operates in even longer pulse QCW mode. The solder layer temperature gradient is steady. A strong driving current may cause the solder layer to experience a significant temperature gradient. The solder layer fully experiences the pulses and the temperature change they cause for pulse widths greater than 900 ns.

There is an obvious temperature cycling on the solder layer. Both thermal fatigue and electromigration are problematic in this situation, particularly when the driving current is large. A typical semiconductor laser operating in short pulse QCW and CW mode has a lower solder joint induced failure under the same driving current, whereas a typical semiconductor laser operating in long pulse QCW mode has a higher failure rate. It is evident that as the heating time increases, the lateral temperature profile in the quantum well alters before stabilizing between 100 and 280 μs. It indicates that transient heat transport has stabilized for the device. Even if the temperature keeps rising with extended heating times, the temperature differential between the emitter and the pitch centers is currently constant. The heat of one emitter would not affect its neighbors when the pulse was less than 200 μs. The neighbors would not be affected by the emitter heat even if one of them stopped working or its heat could not be cooled. This would make the device far more reliable. The single bar CS-packaged 60W 808 nm semiconductor laser was evaluated in various pulses based on the transient numerical simulation (Liu et al., 2015c). The modeling values were smaller than the tested thermal impedance. This is primarily due to the simulation being run with an ideal solder layer. Some solder voids are always present and are unavoidable during production. In modeling, contact thermal impedance and material defects are ignored. There is a comparable trend in the impedance curves derived from modeling and experiment. During the packaging process, certain voids could show up in the solder layer. The heat produced in the quantum well has a tendency to accumulate closer to the voids, raising the local temperature. There could be a substantial modification to the quantum well temperature distribution. This can significantly lower the lifetime and reliability of semiconductor lasers and there is a chance that the spectrum will be significantly expanded. The local temperature increases with the size of the void. However, the local temperature rise will be higher if the solder void is closer to the bar center when the voids are the same size. Quantitative analyses of the impact of void size on local temperature were conducted. The link between the diameter of solder voids and the local temperature rise was established.

7.11. HIGH POWER CS

A high power hard soldered conduction-cooled semiconductor laser bar and its setup contains the cathode, laser bar, mounting substrate, and mounting heat sink that are the four important components (Zhang et al., 2012). First, AuSn solder is used to link the bar to a CuW submount. Next, indium solder or another solder, like SnAgCu, is used to attach the CuW mounting substrate to the copper heat sink. A hard soldered conduction-cooled 40 W 808 nm semiconductor laser package with a 30% filling factor in CW mode serves as the basis for the steady state thermal characteristics. It is assumed that the electrical-optical conversion efficiency is 50%. The mounting substrate thermal resistance plays a major role in the overall effective thermal resistance, therefore choosing the right mounting substrate material is the logical way to enhance the HCS package thermal performance. Analysis shows that using copper-diamond composite material for the mounting substrate rather as CuW clearly lowers the device steady-state temperature. Using a Cu-diamond mounting substrate could lower the heat resistance by roughly 30%. The typical design for a high power semiconductor laser involves cooling from the bottom or rear end. Double-sided cooling has been suggested to increase the efficiency of heat dispersion. The heat produced by the laser bar is one-sided and carry it to the heat sink in the single-sided configuration, which has a low cooling efficiency; in contrast, the heat can be conducted through both the anode and the cathode in the double-sided structure. The effectiveness of thermal dissipation is increased. The thermal behavior of double-sided cooling packing has been investigated through the use of finite element analysis. According to the modeling results, heat is not only directly transferred from the bottom to the heat sink but is also transferred via the cathode, which significantly increases cooling efficiency. The cathode heat dissipation efficiency for the double-sided cooling packaging design can reach 20%. When the drive current is 144 A, a single semiconductor laser bar with a double-side cooler may produce 150 W of power. MaCC is less expensive and suffers less damage from electrochemical erosion than traditional MCC and easy to maintain. High thermal conductivity materials should be employed as the heat sink material in order to achieve good heat dissipation for semiconductor lasers. Copper has long been utilized extensively as a heat sink material in commercial applications because of its high thermal conductivity of about 400 W/(m°K). Higher thermal conductivity heat sink materials have been researched in an effort to enhance heat dissipation. Diamond has been the promising material for creating new heat sink materials because of its high heat conductivity of 2,000 W/(m°K) (Kidalov and Shakhov, 2009). Many types of copper and diamond compound materials have been created, and the copper and diamond compound (CuD) has been suggested as a novel heat sink material in recent years. From 420 to 700 W/(m°K), CuD thermal conductivity has significantly increased, surpassing that of copper heat sinks. Packaging structure is essential for high power laser thermal control. The width of the laser chip, various metallization diffusion barrier structures, and Epi-side Au metallization thicknesses are just a few of the many variables that affect the semiconductor laser thermal characteristics. High power semiconductor laser performance and reliability are impacted by thermal stress that arises during the packaging and operation operations. The mounting substrate and laser chip mismatched coefficients of thermal expansion (CTE) are the primary source of the stress. For high power diode laser packaging, it is ideal to use materials with high thermal conductivities and CTE that are comparable to semiconductor materials like GaAs, InP, and GaN. One of the important issues with high-power diode laser packing is thermal stress. The performance of the semiconductor laser is impacted by thermal stress since it strains the semiconductor and modifies its band structure. The device threshold, wavelength, polarization, and smile can be significantly impacted by thermal stress. The laser device lifetime may be shortened by temperature stress, which could damage the laser chips or bars. The strain brought on by thermal stress divides the heavy-hole E_{HH} and light-hole $_{ELH}$ sub-bands and splits the degeneration at the valence band edge maximum.

7.12. BAND GAP

The band gap of a semiconductor laser will alter due to stress, as seen by a shift in the lasing wavelength. Individual emitter wavelengths of emission may be impacted by thermal stress and packaging-related factors. Due to the wavelength nonuniformity of the various emitters, spectral broadening of laser arrays may occur since they are constructed with many separate emitters. The emitter in the middle of a laser bar usually has a higher thermal stress than the emitter at the surface. The emitters in the middle of the bar exhibit a high variation in wavelength than those near the bar edge. The transition energies shift from lower values for the conduction band (CB) on the copper heat sink to higher values for the CB on the diamond heat spreader. These findings suggest a strong correlation between the output wavelength redshift and blueshift and the compressive and tensile strains, respectively. Which valence band is engaged in the transition determines the polarization of the light that is released. A polarization perpendicular to the quantum well plane (TE mode) results from the transition between electrons and heavy holes, which is facilitated by a compressively strained quantum well. The quantum well plane (TM mode) becomes polarized as a result of the transition between electrons and light holes that occurs in quantum wells under tensile strain. Researchers describe the various transitions based on the polarization states and the Bloch theory. The stress within the quantum well installed on various heat sinks determines the degree of polarization (DoP) (Yan and Zhou, 2019).

7.13. CTE

To compare how the CTE mismatch between the laser device and the heat sink affects the DoP of the bar emitters, an experiment was designed. The laser device is connected to either a newly constructed heat sink or the conventional pure copper heat sink, and it is evaluated at a current that is lower than the threshold current. The silicon carbide diamond in the newly designed heat sink has a copper layer added to both its top and bottom sides. Compared to the conventional pure copper heat sinks, thermal conductivity 400 W/(m°K), this new heat sink has a better match of CTE with the laser device and even high thermal conductivity. The DoP measured in the device installed on the newly designed heat sink, which has reduced thermal stress, changes less than that mounted on a normal copper heat sink, which causes the package to have higher thermal stress. A broader smile may result in a worse beam quality, making laser beam collimation more challenging. Delamination and cracking that occur along the interface are caused by CTE mismatches between the metallization, die attachment, and copper heat sink. When the applied stress exceeds the laser material strength, bar cracking may result from the stress in the materials caused by the CTE mismatch between the diode and its mounting substrate. The internal forces that nearby particles of a continuous material apply to one another are expressed physically as stress. Since the heat sink (copper) has a larger CTE than the laser array, which is composed of GaAs material, thermal stress is an inherent issue with the semiconductor laser. There is a 14 μm contraction difference between the Cu heat sink and laser array along the length of the laser array for a typical 10 mm long laser array, due to the CTE difference of 11×10^{-6}/°C between copper and GaAs material and the temperature difference of 131°C between the indium solder freezing temperature and room temperature. The primary cause of the laser diode array thermal stress is the CTE mismatch. Thermal stress is produced during the cooling phase of the bar bonding process as a result of CTE mismatch.

One needs to employ a compliant solder, like indium, that can relieve thermal stress by plastic deformation when there is a significant CTE mismatch. A perfect match between the CTE of the laser chip and the CTE of the heat sink in the laser package would be ideal. The availability of such a heat sink is very difficult. Depending on the heat sink composition and package design, there is always a CTE mismatch between the laser chip and the heat sink in a laser package, although to varying degrees. Simulations are performed on a laser bar packaged on a heat sink by adjusting the expansion coefficient of the heat sink around the GaAs CTE value in order to understand the impact of CTE mismatch on the residual stresses in the laser bar. Two types of solders are frequently used in semiconductor laser packaging: AuSn solder, which is a hard solder, and Indium solder, which is a soft solder. Compared to AuSn, indium has a substantially lower yield strength and is more susceptible to plastic deformation under stress. The solder plastic elongation during a heat cycle can help reduce stress when using Indium. The maximum tension placed on the laser device is reduced by thickening the indium junction. Thermal fatigue, electro-migration, and thermoelectro-migration of the indium solder in the laser package significantly diminish the reliability of the semiconductor laser linked with indium solder, despite the fact that indium can be utilized to lessen thermal stress in high power semiconductor lasers. These problems can be resolved with AuSn, which is stronger than indium. The AuSn utilized as the bonding medium permits minimal stress and strain compensation. When using AuSn solder, a buffer layer (CuW) should be added to lessen the heat stress on the laser chip. Like Indium, AuSn thickness can have an impact on the laser package interfacial stress. When the solder layer thickness is increased from 20 to 60 μm, the shear stresses at the substrate-solder and chip-solder interfaces are decreased by around 40%. AlN, AlSiC, CuW, and Al_2O_3 are commonly utilized as mounting substrates in laser diode packing. AlN and BeO are dielectric materials used for electrical insulation, whereas the CuW composite is usually utilized for heat spreading. The mounting substrate has a direct impact on the laser device thermal stress. Numerical analysis has been done to determine how the substrate affects the maximum interfacial shear stress and normal stress, assuming that the maximum temperature difference is between room temperature and the soldering temperature of 300°C. When bonding the GaAs chip to a copper or diamond substrate, the calculated interfacial shear stresses are 387 and 168 MPa, respectively (Fabis, 1999). During the cooling stage of the soldering process, a plastic deformation is caused in the solder because the estimated interfacial shear stress is greater than the yield stress of the gold-tin solder. The pressures in the solder layer are redistributed by this plastic deformation, which ultimately lowers the GaAs chip stresses. The Al_2O_3 substrate exhibits the least amount of thermally induced shear stress in GaAs laser chips, but the Cu substrate exhibits more stress. A unique sandwich design for the mounting substrate has been created in order to increase thermal conductivity and improve CTE fit with the chip. A layer of copper is positioned in the center of the sandwich mounting substrate, while the top and bottom of the sandwich are made of a substrate material that has better CTE match and good thermal conductivity than copper. By selecting the appropriate copper layer thickness, the sandwich substrate CTE can be reached. The sandwich CTE varies significantly when the copper thickness varies from 25 to 100 μm.

The layer system of 300 μm diamond, 50 μm copper, and 300 μm diamond has the closest CTE match to GaAs (Fabis, 2002). The estimated effective thermal conductivity of this sandwich substrate is almost 1,500 W/(m°K). The thermal stress brought on by the CTE mismatch between the laser chip and the mounting substrate determines the maximum energy transition values in laser bars. It is evident that the maximum energy transition value shifts the least on 50 μm-DCD substrates with the best CTE match to the laser chip, whereas the maximum energy transition value shifts the most on laser bars mounted on bare copper mounting substrates with the largest CTE mismatch.

7.14. FABRICATION

The goal of fabricating silicon carbide diamond (ScD) is to decrease heat resistance and CTE. The copper on the ScD body is brazed on both the top and bottom surfaces. ScD is less expensive to manufacture than high-pressure diamond composites. ScD material is used to design and build the initial heat sink samples. A CTE of 7–8 ppm/K and a thermal resistance of 0.7 K/W have been attained by adding top and bottom copper layers of accurate thickness. Microphotoluminescence spectroscopy (μ-PL) can be used to characterize the thermal stress in the laser package. The laser package lattice strain or stress determines the exact wavelength of the PL signal. An increase in stress at the laser edges is shown by the longer wavelength. At the edges, the wavelength change is around 2 nm. A laser bar positioned on the ScD heat sink, however, displays a flat line. The results demonstrate that using a ScD mounting substrate can significantly lessen the stress created in the laser bar. A thin layer of copper is electroplated onto the molybdenum surface. The low thermal resistance is caused by the thin copper layer acting as a head spreader. In comparison to the molybdenum layer, the thin copper layer CTE effect is minimal because of its thinness. Bulk molybdenum has a CTE of about 5 ppm/K, which is somewhat less than that of GaAs. The symmetrical nature of the sandwich design prevents the heat sink from bending when the temperature changes. The CTE and thermal resistance of the molybdenum-copper-molybdenum-structured heat sink are less than 9 ppm/K and 1 K/W, respectively, than those of the conventional copper heat sink. For mounting substrates, composite materials containing copper filled with diamond fillers are utilized to provide a high thermal conductivity and good CTE fit with the chip. Powder metallurgy has been used to create copper-diamond filler (CD) composites. By altering the volume of diamond fillers, the composite thermal conductivity and CTE can be altered. While the CTE decreases as the diamond content increases, the thermal conductivity increases as the diamond content increases. When the CTE of the copper-diamond composites is the same, the high purity diamond with the bulk thermal conductivity of 2,000 W/(m°K) exhibits superior thermal behavior in the composite materials compared to the diamond with the bulk thermal conductivity of 1,000 W/(m°K). The high purity diamond composite mounting substrate shows around 50% higher heat conductivity than that with the diamond of 1,000 W/(m°K) when the mounting substrate matches the chip CTE.

7.15. APPLICATIONS

The semiconductor lasers are gaining importance as light sources in a variety of industries, including medical, high-power solid-state lasers, direct material processing, and fiber laser pumping. However, the primary problem preventing them from being used in more applications is the low beam quality. Beam shape and optical design are crucial for increasing the uses of semiconductor lasers. The waveguide characteristics of the semiconductor laser active region are to responsible for the low beam quality. Several common mathematical models, including the Gauss model and Lorentz model have been investigated to characterize the source field of a single emitter semiconductor laser. When the beam is converted by passive optical parts like lenses or mirrors, this beam-parameter product stays constant in the optimal situation, that is, when there are no aberrations. The diffraction limit is the lowest value for the beam-parameter product. Fast axis collimators (FACs) are used to collimate laser bars in a laser stack (Lutz and Poyet, 2014). The collimated beam pointing error or CBPE affects the laser stack. Three elements contribute to the laser stack CBPE. The first is the laser bar smile. The collimated beam quality of the laser bars is not considerably diminished if they have a moderate smile of less than 1 μm in the laser stack. The FAC installation error is the second factor. The stack installation error during the laser stack assembly procedure is the third factor. The micro-aspheric lens is called the FAC, and its position has a significant impact on the laser bar far-field intensity distribution. It is evident that even minor FAC installation problems might lower the quality of the collimated beam.

7.16. ERRORS

Errors in the laser stack installation can potentially result in CBPE during the assembly phase. It indicates that there are variations in the positions of each laser bar within the laser stack during packaging, such as tilting or displacement of the laser bars along the fast or slow axis. Another crucial step is to minimize installation errors during stack packing. In order to maintain a consistent bar pointing direction, fast axis collimation components are first introduced. Each bar location is then adjusted from a vertical and horizontal orientation utilizing sophisticated real-time monitoring technology to guarantee precise positioning. This method may effectively fix the BPE in a laser stack. The poor beam quality of semiconductor lasers should be corrected prior to application. In certain applications, fiber delivery of the laser beam is necessary. Improving the beam quality of semiconductor lasers and fiber coupling involves some crucial processes. They are fiber coupling, beam shaping, and beam collimation. A single emitter laser active zone is moderate, measuring 1 μm along the fast axis and 100–200 μm along the slow axis, respectively (Petrie and Toth, 2018). The slow axis FWHM divergence angle is 8°. A single emitter laser can be directly connected into a fiber with a diameter of 100–200 μm following beam collimation in the fast axis. The wedge-shaped input end of the fiber collimates the divergence angle along the fast axis. The fiber wedge-shaped tip acts as a collimator. The beam is linked into the fiber following the wedge collimation. However, the wedge shape is necessary to achieve high fiber coupling efficiency, and it is challenging to produce accurately and with high yield. This restricts how this technology can be used. A collimator can collimate the beam prior to fiber coupling, reducing component requirements during processing. A cylinder lens is always employed to collimate the beam along the fast axis because of its small size. The self-focusing lens is a gradient index lens, meaning that as it moves radially, its inner index gradually decreases. One typical optical component for focusing light beams is the graded-index lens. The semiconductor laser beams have been coupled using it. A single semiconductor laser fiber coupling system uses a graded-index lens. Three crucial optical components in this system are fiber, self-focusing lens, and FAC. The self-focusing lens is used after the FAC to focus the beam. The FAC is a small cylinder lens that is bonded in front of the semiconductor laser. For a single emitter laser, the fiber connection is simple due to its small beam size. However, the beam quality of a semiconductor laser bar is inferior than that of a single emitter laser, and its beam size is 10 mm along the slow axis. The beam shaping along the slow axis is also required in addition to beam collimation along the fast axis. In fast axis, there are two steps involved in semiconductor laser collimation. A semiconductor laser's beam is first collimated along the fast axis and subsequently along the slow axis. A semiconductor laser source can be regarded as a point source due to its small size on the fast axis. Geometric theory states that if the right collimation lens is made, it can be collimated. Following collimation, a fiber array bundle can be used to link the semiconductor laser bar beam. Another optical shaping component is the monolithic micro-cylinder system. It is made up of N micro-cylinders, each of which is tilted along the slow axis. Quartz is typically used to manufacture the system. The entire system is referred to as the beam transformation system (BTS), and the optical components and rapid collimation lens are often consolidated into an integrator. The design of the focusing optical system should take into account the fiber numerical aperture (NA) requirements. After beam shaping, the beam along the slow axis needs typically be re-collimated, and a cylinder lens is always employed in this process.

7.17. PACKAGING PROCESS

The fundamental packing procedure for high power semiconductor lasers has not altered much in spite of many advancements in production. (1). Incoming material inspection; (2). raw material cleaning; (3). Metallization; (4). solder deposition; (5). die bonding; (6). wire bonding; (7). Assembly; (8). Screening; (9). before burn-in (BBI); (10). burn-in (BI); (11). after burn-in (ABI); and (12). final inspection are all steps in the fabrication process of open packages. An open package, optics, and a fiber are used in fiber-coupled modules to couple light into the fiber. Incoming material inspection is the initial stage of the packing process that separates the undesired materials for cost and quality control. All raw materials are cleaned using various cleaning techniques to get rid of oxidation and contaminants. Following cleaning, layers of solder and metallization are applied to the soldering target. The technique of joining a laser chip to a heat sink is called die bonding (Zhou et al., 2022). In certain configurations, the cathode and laser chip are connected by wire bonding. The assembly step involves putting the laser device together using many packaging components. Following assembly, screening is done to look for laser defects like misalignment, solder overflow, soldering voids, and facet defects. The diode laser performances are then evaluated by BBI. Burn-in is used in all products for quality control as a necessary and practical method to quickly screen unqualified products. ABI tests characterize the diode laser performance again. The last stage to ensure that there are no more problems with the product is final inspection. The popular and crucial stage of quality and yield control is incoming material inspection (IMI), also known as receiving inspection. IMI is responsible for ensuring that the entering materials fulfill the manufacturing criteria and material specifications. It eliminates losses and lowers manufacturing costs. It is crucial for preserving the uniformity of incoming raw materials and enhancing the reliability and quality of the final products. IMI also collects data about their properties. To ensure there is no damage to the package or its components, the received package is visually inspected to get an idea of its state. The components and substrates are inspected and their dimensions are measured using profile projectors. Surface roughness (some areas) and flatness are measured using surface profiling equipment. Both visual inspection and microscopy are used to examine the arriving materials, including mounting substrates, coolers, laser chips, electrodes, and insulators. This procedure identifies all surface defects, usually includes scratches, contaminations, and chipping. The laser chip is bonded to the mounting substrate, which can be a copper-tungsten (CuW) substrate, copper block, ceramic substrate, or micro-channel cooler (MCC) (Liu et al., 2015c). Achieving a high-quality die bonding depends on the surface quality of the critical region where the laser chip is bonded. The critical region needs a stringent inspection standard than the others, and it requires more attention. For solder materials, melting temperature is a crucial factor. The melting temperature of solder materials is inspected and confirmed for contamination, oxidation, and material composition throughout the actual manufacturing process. By merely watching the melting and freezing process in a reflow oven with a specific preset temperature profile, one can estimate the melting temperature. The differential scanning calorimetry (DSC) method provides a precise melting temperature (Spink, 2008). DSC is a thermoanalytical method that measures the variation in heat needed to raise a sample and reference temperature as a function of temperature. The fundamental idea behind determining the melting point of solder materials is that, because the sample absorbs heat during the endothermic phase transition from solid to liquid, more heat will be required than the reference to keep both at the same temperature when the solder material sample goes through this phase transition. The endothermic event's onset temperature is used to calculate the melting temperature.

7.18. METALLIZATION

To accomplish high-quality diode laser packaging, the metallization must adhere to the body material, whether it be the heat sink or the mounting substrate. The laser chip in high power semiconductor laser packing is soldered to a mounting substrate, which is joined to a heat sink or cooler by metallurgical bonding. The heat sink and mounting substrate materials must be solderable. The metallization of the mounting substrates and heat sink material surfaces is necessary. High power diode laser performance and quality are strongly impacted by metallization adhesion. The cross-cut and tape test are used to demonstrate the adhesion capability on the surface of a heat sink that has been metallized with nickel and gold. When sticky tape is applied and peeled at the cross-cut area, metal with good adhesion does not peel off or separate. For die bonding, the solder thin-film layer thickness and the metallization layer thickness must be kept within a specific range. An X-ray fluorescence (XRF) tester can be used to determine the coating layer thickness and composition (Fiorini et al., 2002). The emission of distinctive secondary or fluorescent X-rays from a material excited by high-energy X-ray bombardment is known as XRF. The energy and intensity of the X-ray emission are used to calculate the coating thickness. The observed thickness varies between 10Å and several micrometers. The value of the contact angle θ is typically used to evaluate the wettability. Because the metal surface condition has a significant impact, the wettability of the same metal substance might vary greatly. The wettability of the metal submount can be significantly impacted by thin coatings of oxide or organic pollutants on its surface. The wettability can be assessed using the contact heating method. Incoming components should be cleaned as per the packaging process specifications to remove oxidation and contamination from the material surface and enhance solderability. Eliminating organics, dust particles, and other contaminants that degrade device quality and impact production output is crucial. The fabrication process, shipping, and storage are the primary sources of particle contamination. Oil or grease from the machining process, polymers, or photo resist are examples of organic contamination. Oxide may result by natural oxidation; for instance, copper and indium can readily oxidize at room temperature. Chemical oxidation by other substances may potentially be the cause. Initially, ethyl alcohol ultrasonic cleaning is used to get rid of dust and other debris. The surface of the mounting substrate is then chemically cleaned of organic impurities like particles, grease, or silica gel. The plasma cleaning process eliminates the oxide layer. A chemical solution is used to remove the heavy metal and ion impurities. The three primary cleaning techniques utilized in the semiconductor laser packing process are ultrasonic, chemical, and plasma cleaning. Every component utilized in the packaging of laser devices, including mounting substrates, heat sinks, coolers, insulators, solder, and electrodes needs to be cleaned. Different tools and procedures should be utilized to clean up the incoming components because different materials used in high power diode laser packaging have varying characteristics, and coating layers. It can be sufficient to clean some parts with acetone and then clean them with isopropyl alcohol. A thorough cleaning is required for the majority of the components and materials. Cavitation bubbles created by high frequency pressure (sound) waves are used in ultrasonic cleaning to agitate or stir a liquid (Yasui, 2018). When impurities adhere to materials such as metals, polymers, glass, rubber, and ceramics, the agitation creates strong forces. This action enters cracks and blind spots. The goal is to completely eliminate any contamination that is firmly adhered to or imbedded in solid surfaces. Depending on the kind of pollution, either water or other solvents like acetone and anhydrous ethanol might be employed. To clean sensitive parts, ultrasonic cleaning uses ultrasound (20 and 400 kHz) and a suitable cleaning solvent, occasionally just regular tap water. Cleaning typically takes a few minutes, but depending on the component being cleaned, it may take up to 20 minutes. It may not be necessary to disassemble the component before using ultrasonic cleaning, which can be applied to a variety of materials.

Organic, oxide, and metallic impurities are commonly found on the parts used in high power laser packaging. These contaminants can originate from a variety of lubricants, natural or chemical oxidation, corroding metal, the machining process, and hard water deposits. Alkali cleaning is the first stage. Eliminating insoluble organic pollutants like silica gel, oil, and dust particles is the aim of this stage. It is possible to utilize several chemical solutions. One often used solution is the 5:1:1 $H_2O:H_2O_2:NH_4OH$ combination. Rinsing is the second stage, which removes the deposits from the components and uses deionized water to wash away the compounds and alkali. Acid cleaning is the third stage. The type of material to be cleaned and the pollutants to be eliminated determine which acid is best. Ionic, heavy metal, and indium oxide layer pollutants can be removed using a hydrochloric acid (HCl) solution, whereas a silicon dioxide layer can be removed successfully using a hydrofluoric acid (HF) solution. Rinsing, the fourth stage, is comparable to the second but is done to remove the acid solution. Depending on the nature and purpose of the items to be cleaned, the rinse liquid in this phase can also be anhydrous ethanol, acetone, or deionized water. Since the cleaned metal is now extremely reactive, a protective layer must be applied to stop corrosion. In order to create an oxide layer on the metal, the fifth step involves reactivating the metal with an oxidizing agent, often an alkali. The metal underneath is shielded by this layer, which cannot oxidize any further.

7.19. PLASMA CLEANING

Plasma cleaning (Figure 7.3) is used to get rid of extremely thin coatings that are left over after traditional cleaning, particularly hydrocarbonats and oxides (Aronsson et al., 1997). Plasma cleaning increases the surface wettability, eliminates organic contaminants from the surface, improves adhesion on polymers, glass, ceramics, and metal, and usually doubles the shear strengths. Plasma cleaning is the process of using an intense plasma or dielectric barrier discharge plasma produced from gaseous species to remove pollutants and impurities from surfaces. Air and hydrogen/nitrogen mixes, as well as gases like argon and oxygen, are employed. The low-pressure gas is ionized using high frequency voltages to produce the plasma. Selecting the appropriate plasma gas is crucial since different gases react and remove impurities in different ways. Plasma is an efficient, cost-effective, and ecologically friendly technique for cleaning when oxygen gas is employed. The majority of organic bonds of surface pollutants can be effectively cleaved by vacuum ultraviolet light. This aids in the disintegration of pollutants with a high molecular weight. The oxygen species produced in the plasma provide a secondary cleaning function. These species produce smaller molecular weight hydrocarbons, CO, H_2O, and CO_2 when they react with organic pollutants. During processing, these compounds are removed from the chamber due to their comparatively high vapor pressures.

The surface that is produced is extremely clean. Organic residues and coatings can be effectively removed by oxygen plasma.

Figure 7.3. *Plasma cleaning.*

Source: https://upload.wikimedia.org/wikipedia/commons/f/fc/Oxygen_Plasma_Cleaning.JPG, Creative Commons License.

The use of inert gases like argon or helium is used if the component to be treated is made of easily oxidized materials like copper or silver. Organic pollutants can be broken down by the plasma-activated atoms and ions, which act like a molecular sandblast. During processing, these impurities are evaporated and removed from the chamber. Before soldering, argon plasma cleaning can be used to eliminate the oxidation on the indium layer. Contact angle measurements can be used to determine whether organic removal is complete. Water will have a large contact angle with the surface when there is an organic contamination present. The contact angle will decrease following the elimination of the contaminant. When a soldering procedure is required, mounting substrates, heat sinks, and other components must be metalized. Metallization acts as a wetting layer and diffusion barrier between the solder and substrate materials. It improves the device reliability, lowers contact resistance, and stops oxidation or atmospheric moisture. Prior to metallization, the target surface flatness and roughness must be carefully managed because undesired surface conditions could result in pinholes or defects on the coating layer.

CHAPTER 8

ULTRAFAST SEMICONDUCTOR LASERS

Contents

8.1. INTRODUCTION

Short optical pulses can be produced from semiconductor diode lasers in several ways. These comprise gain switching, Q switching, and mode locking. Gain switching depends on the drive current being modulated in order to switch the optical gain. An external cavity is not required for gain switching. Short pulses can be obtained by modulating any type of diode laser, distributed feedback

(DFB), Fabry-Perot (FP) lasers (Figure 8.1), or distributed Bragg reflector (DBR), provided that the modulation power is high (Kojima and Kyuma, 1990).

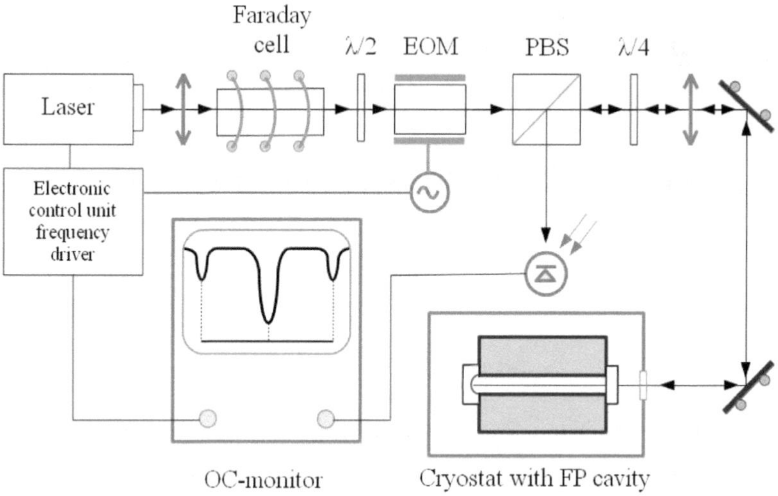

Figure 8.1. *Fabry-Perot (FP) laser.*

Source: Kuvshinskii, M. V., Oreshkin, S. I., Popov, S. M., Rudenko, V. N., Yudin, I. S., Azarova, V. V., & Blagov, S. V. (2019). Tests of cryogenic fabry–Perot cavity with mirrors on different substrates. Applied Sciences, 9(2), 230. https://doi.org/10.3390/app9020230, Creative Commons License.

Gain switching has some benefit over mode locking in that the frequency of the modulation current can alter the repetition rate. Compared to gain switching, mode locking produces shorter pulses. Compared to the electrical pulse, the optical pulse produced by gain switching is shorter. Many studies have been conducted on the gain-switching phenomenon. A set of nonlinear rate equations of a diode laser must be studied in order to comprehend the dynamics of gain-switched pulses in diode lasers. In a distributed feedback (DFB) diode laser, the wavelength of the CW laser is 1549 nm. This DFB laser has a 10 mA threshold.

8.2. BIAS CURRENT

At a bias current of about 20 mA, the coupled power into the fiber is greater than 1 mW. A synthesizer provides the modulation signal. Approximately 20 dBm is the modulation power and the output pulse width is measured using the autocorrelator. The DC bias current has a significant impact on the gain switching. It is evident that gain-switched pulses are unstable when the bias is too low and broad when the bias current is too large. To obtain the shortest pulses, the DC bias current must be optimized. It is important to note that as the modulation frequency (repetition rate) varies, so does the ideal bias current. The laser carrier dynamics must increase in order to react to the modulation signal as the modulation frequency rises. A larger bias results in a faster diode laser modulation speed. For this reason, when the modulation signal repetition rate rises, the bias current must be raised to obtain the shortest pulses. Because carrier dynamics in the laser are quick at low repetition rates, the optimal DC bias does not rise with repetition rate. The optimal DC bias current rises linearly with the modulation frequency at large repetition rates. A higher modulation power results in a higher peak carrier density and a narrow pulse width. As the modulation power rises, the gain-switched pulse peak power rises as well, eventually reaching saturation. It is known that carrier density affects the laser refractive index. A gain-switched pulse emission causes a significant fluctuation in the carrier concentration. The refractive index should therefore change as well and compared to CW operation; the spectral width is wider with gain switching. The basic idea is that by injecting the pulse into a fiber where the blue-shifted light moves more quickly than the red-shifted light, the pulse will shorten since the front of the pulse is blue-shifted and the back is red-shifted. A dispersion compensating fiber, or DCF, is an example of this sort. The models and experiments demonstrate that a semiconductor laser may be used to readily generate gain-switched pulses at different repetition rates.

8.3. LASER PULSES

One method for producing short laser pulses is mode locking. By modulating the laser gain or loss at the inverse of the optical cavity roundtrip time, this approach locks the longitudinal mode phases. Mode-locked semiconductor lasers have been demonstrated many times using the external cavity setup. The gain element is an AlGaAs laser chip (Figure 8.2) (Mobini et al., 2022). On the side with the optical cavity, there is an antireflection coating. Oxygen ion bombardment produces a saturable absorber on one facet. The depth of implantation is roughly 10 mm. The cavity round-trip time is 1.2 ns. The current is modulated at a frequency that corresponds to the inverse round-trip duration, and the laser is DC bias. Stable mode-locked operation requires precise laser cavity tuning. A pulse with a 560-fs pulse width and a repetition rate of about 625 MHz is produced by using proton bombardment as an absorber. For a variety of fiber transmission systems, mode-locked semiconductor lasers with transmission data rates of at least 10 Gb/s are crucial. These consist of clock recovery and soliton creation in extremely fast transmission systems. Since they are more stable than external cavity designs, single-chip, mode-locked lasers are preferred. The InGaAsP/InP material system is used to fabricate the laser.

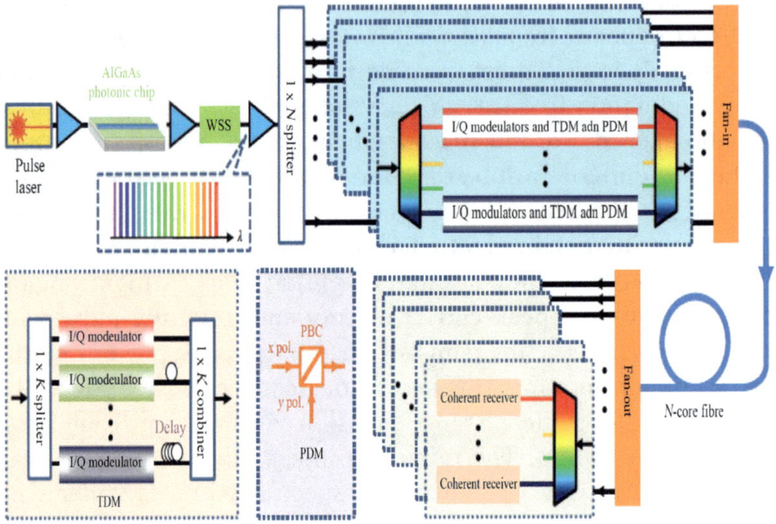

Figure 8.2. *AlGaAs laser chip.*

Source: Mobini, E., Espinosa, D. H. G., Vyas, K., & Dolgaleva, K. (2022). AlGaAs nonlinear integrated photonics. Micromachines, 13(7), 991. https://doi.org/10.3390/mi13070991, Creative Commons License.

At the center of the optical cavity is a saturable absorber. When the mode-locked pulses from opposing directions arrive at the saturable absorber simultaneously, the saturation is increased and a short pulse may result. This is called ad Colliding pulse mode locking (CPM). In the CPM arrangement, the round-trip time is t = nL/c, where c is the speed of light in vacuum, n is the index, and L is the linear cavity length. Ultrashort transform-limited pulses are produced in dye lasers using the CPM technique.

CPM lasers (Figure 8.3) use the coherent interaction of two counter-propagating pulses colliding at the saturable absorber to create short pulses rather than a single pulse traveling through the absorber at once. Since this operation has the lowest loss at the saturable absorber, the laser works with two counter-propagating pulses colliding at the same time. The pulses are stabilized and shortened by the absorber increased effective saturation cross-section. There are four gain sections on the laser. The optical cavity has two portions in the center and two sections at either end. A waveguide layer connects these portions, allowing light to flow through them while being pushed with enough current to provide gain. In the middle is the saturable absorber.

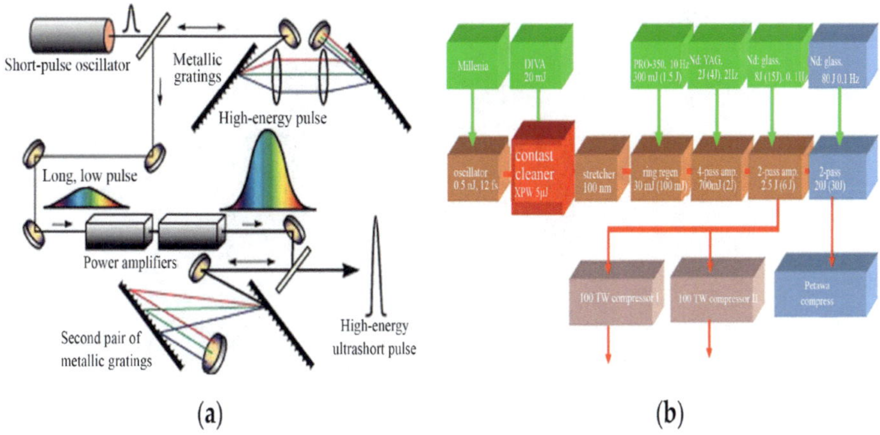

Figure 8.3. *CPM laser.*

Source: Chvykov, V. (2024). High-power lasers. Encyclopedia, 4(3), 1236-1249. https://doi.org/10.3390/encyclopedia4030080, Creative Commons License.

Every portion has both RF and DC contacts. The saturable absorber part receives the modulation. The waveguide runs along the chip beneath the active area. Through this waveguide, the light produced in the active zone is reflected back. The cleaved facets create the optical cavity. The saturable absorber part is about 150 μm long, and the four active gain sections are around 350 μm long. The chip is around 8.6 mm long. The mode-locked pulses have an estimated round-trip time of 0.1 ns, or a repetition rate of roughly 10 GHz. Around 1560 nm is when the laser emits. When connected to a single-mode fiber, the laser threshold current is 65 mA, and its differential efficiency is 0.04 mW/mA (Rao et al., 2013).

8.4. MANUFACTURE

The following procedures are involved in the device manufacture. First, a (100)-oriented InP substrate is used to produce the waveguide area or region, a p-InP cladding region, a thin InP region, the multiquantum well active region with four quantum wells, and a thin p-InGaAs region. The active area of the multiquantum well is made up of barrier regions that are 80-A thick and active region wells that are 100-A thick (InGaAs). The saturated absorber zone also contains quantum wells. The wafer is patterned with dielectric, the thin p-InGaAs layer is removed by etching, and in certain areas, the wafer is etched past the active region. The wafer is then produced with a second set of layers that include a p-InP cladding layer and a p-InGaAs contact layer.

Figure 8.4. *Laser chip.*

Source: https://upload.wikimedia.org/wikipedia/commons/d/d5/4-fach-NAND-C10.JPG, Creative Commons License.

The laser chip (Figure 8.4) is created by processing the wafer using conventional photolithography, metallization, and cleaving techniques. For testing, the submount that the laser is bonded to is placed on top of a thermo-electric regulated heat sink. By modulating the laser at frequencies close to 10 GHz, the mode-locking behavior of the laser is examined. The autocorrelator is used to measure the pulse width. For a range of bias currents, the pulse width is small, and it gets wider after that. The change in round-trip time with injection current may account for the refractive index dependence on the current. It was discovered that the saturable absorber pulse width did not change with RF modulation power. There have been reports of mode-locked lasers with a short cavity and a colliding pulse system that run at high speed. A saturable absorber zone along the optical cavity is surrounded by two gain portions. The InGaAsP/InP material system was used to produce the laser, which has an emission wavelength of about 1520 nm. To lower the capacitance, the Fe:InP semi-insulating layer is required. On the upper surface are both laser contacts. We can create lasers with different cavity lengths and the speed of the short cavity lasers was higher. The shortest lasers had a round-trip time of 2.87 ps and a linear length of 534 mm. At 350 GHz, transform-limited pulses with a pulse width of 640 fs were produced. The 350 GHz pulses measured optical spectra reveal an FWHM spectral width of 4 nm. This translates to a time-bandwidth product of 0.32 that is almost transform-limited. In passive mode-locked lasers, the absorber must saturate more quickly than the gain in order for pulse shaping to be successful. The leading edge of the pulses is sharpened as a result, while the trailing edge is diminished as the gain reaches saturation. As the pulse travels through the laser cavity several times, the pulse shortening process keeps going. Dispersive effects eventually stop the pulse from getting any narrower. The saturable absorber physical length should be less than the optical pulses in order to sharpen them. The absorber physical dimensions and the dispersion from the active zone set a minimum pulse width limit. The pulse width may be decreased by optimizing the absorber length, saturation property, and low dispersion. Two-level amplitude or intensity modulated signals are used for transmission in the majority of optical transmission systems. The high intensity (S_1) and low intensity (S_0) of the transmitted signal determine the 1s and 0s for digital transmission systems. This process, known as pulse code modulation (PCM), involves modulating the laser with the current to transform a digital current pulse pattern into an optical signal pattern. High frequency modulation, or the laser modulation characteristics must be high for high transmission systems. The frequency at which the modulated light power is half of its value at low frequencies is sometimes referred to as a 3dB bandwidth (Yabre, 1996). This 3dB bandwidth must be high enough, it must be greater than 15 GHz for 20 Gb/s PCM.

8.5. MODE-LOCKED LASERS

Novel semiconductor lasers like nanostructure lasers and small mode-locked lasers are the result of advancements in materials technologies. The transmission optical fiber contains a core that is roughly 10 μm in diameter and a cladding layer that is roughly 100 μm in diameter. Because the core region has a slightly higher index, light can be directed there. A polymer jacket that surrounds the cladding further offers strength and protection. The fiber core of the fiber laser is doped. When pumped by an external pump laser, the doping material like Er, Yb, Nd, Ho, Tm, etc., supplies the gain required for laser operation. In rare-earth doped fiber (optical amplification), the concentration of rare-earth dopants can vary significantly. In multicomponent glasses, it can reach approximately 1000 parts per million (ppm), whereas in erbium-doped fibers, it can be as low as less than 100 ppm. It may be a few ppm in erbium-doped fibers used for distributed amplification. Several techniques can be used to produce optical fibers. These include outside vapor deposition (OVD), axial vapor deposition (VAD), modified chemical vapor deposition (MCVD), and plasma-assisted chemical vapor deposition (PCVD). For silica fiber, rare-earth dopants are frequently added while the fiber is still growing. The production of ordinary low-loss transmission fiber is closely related to the production of rare-earth-doped fiber. Rare-earth dopant levels are typically around 100 ppm. Rare earth-doped glasses share many of the same issues as lasers based on atomic or molecular transitions in gaseous media, including population inversion. This is common with CO_2 lasers, Ar^+ lasers, and He-Ne lasers. Er, Yb, Nd, Tm, Ho, and other rare-earth dopants are frequently employed in glass fibers. Each of these materials has a unique energy level, and if the population of atoms in the higher energy level is greater than that in the lower energy level, the optical transitions between these levels cause light to be absorbed and emitted as well as, eventually, laser action. The use of er-doped glass amplifiers in optical fiber communication systems has been thoroughly investigated. In 1965, the ^{4I} to ^{4}transition was used to show the first Er-doped glass laser (Stone and Bray, 1996). The laser wavelength was close to 1500 nm. Fiber amplifier region of interest for optical communication applications is 1550 nm. Absorption of 980 nm or 1480 nm light from a pump laser is a popular method for exciting Er ions to the excited state. For high-power laser sources and amplifiers in the 1100 nm spectrum, Yb-doped glass amplifiers have been thoroughly investigated. There are two energy-level manifolds for the Yb ion in silica. They are the excited-state (^{2}) and ground-state (^{2}) manifolds, respectively. Strong homogeneous and inhomogeneous broadening prevents the transitions between the sublevels for Yb ions in a glass at room temperature from being fully resolved. In 1988, the (^{2}) to (^{4}) transition was used to show the Yb-doped silicate glass laser. The laser wavelength was close to 1100 nm.

8.6. GLASS LASER

In 1961, a glass laser doped with Nd was reported. From ultraviolet to near-infrared wavelengths, Nd exhibits a significant absorption spectrum. Emission bands in the 1400–2700 nm range are of primary importance for Tm, especially at energies where Er and Nd gaps exist. The typical glasses of Tm doping include silica, phosphates, tellurites, and fluorides. The glass host affects the emission intensities of various transitions. The transmission optical fiber consists of a core that is roughly 10 μm in diameter and a cladding that is roughly 100 μm in diameter. The index rises when Ge is added to SiO_2.

Figure 8.5. *Optical fiber.*

Source: https://upload.wikimedia.org/wikipedia/commons/4/49/Fibreoptic.jpg, Creative Commons License.

Optical fiber (Figure 8.5) production is crucial since nonlinear optical fibers and dispersion-tailored fibers without rare-earth atom doping are used in a number of pulse compression techniques. The core of the fiber used in fiber amplifiers is doped with rare-earth dopants, which makes it different from ordinary optical fibers. Usually, these are Er, Yb, Nd, Er-Yb, and so on. Adding feedback to a fiber amplifier creates a fiber laser. Rare-earth doped lasers have been reported to produce mode-locked pulses through the use of both active and passive mode-locking techniques. A modulator that modifies the optical loss or phase in the optical cavity is typically present in the laser cavity for active mode locking. The mode-locking properties are determined by the saturation parameters of the saturation mechanism, specifically the time constant and the saturation power. In certain situations, stable mode-locked pulses are produced by combining active and passive mode locking. For high bit rate optical communication, ultra-short pulses with a high repetition rate and a wavelength of about 1550 nm must be generated. Such pulses could originate from actively mode-locked Er-doped fiber lasers, however unless a harmonic mode-locking technique is employed, the pulse repetition rate is often constrained by the bandwidth of the active mode locker utilized in the cavity. Active mode locking usually uses a fiber laser in a ring cavity. An optical loss modulator, usually a $LiNbO_3$ modulator, is present in the ring cavity. A sequence of pulses at the modulation frequency are produced when the modulation frequency is an integer multiple of the inverse of the cavity round-trip time. When the modulation frequency is set at (n + 1/p)fc, where fc is the longitudinal mode spacing or fundamental cavity frequency of the fiber laser ring and n, p are integers, an optical pulse train with a repetition rate of (np + 1)fc can be produced. This technique, known as rational harmonic mode locking, has garnered a lot of interest. The experimental findings of several researchers in actively rational harmonic mode-locked fiber lasers have been presented. For many applications, mode-locked operating stability is crucial. One crucial factor to take into account for the actively mode-locked fiber laser commercialization is pulse train stabilization. The mode-locked state cannot be maintained for very long without additional treatment of the fiber laser because of the cavity length drift brought on by different environmental changes (temperature). In order to solve this issue, researchers have shown how to dynamically modify the cavity length in order to stabilize the laser. A phase-locked loop based on the regenerative form of feedback has also been used to stabilize the fiber laser. Output (a portion) is taken out and transformed into an electrical signal for investigations. After being amplified and sent through a narrow-band high Q band-pass filter, the signal is fed back into the modulator RF port, closing the loop. At the band-pass filter center frequency, the laser generated a pulse train. The laser output falls under the harmonic mode-locking category since it operates at the same frequency as the RF driving frequency.

8.7. LOOP

By employing a feedback loop to regulate cavity length, stable mode-locked operation has been demonstrated. A piezoelectric controller (PZT) is employed in the experiment within a fiber ring laser optical cavity. For harmonic mode locking, the ring laser has a $LiNbO_3$ modulator and EDFA (Zhou and Zhang, 2010). Even a small difference between the external modulator frequency and the multiples of the cavity frequency causes the mode locking to occur. The RF spectrum of the mode-locked pulse wave shows differential frequency. The slow periodic modulation in the pulse temporal position is caused by the frequency difference. After the photodiode and frequency counter have detected the deviation frequency, the computer signals the PZT driver to lock the frequency. This is referred to as asynchronous mode locking. A modified method employing a phase-locked loop (PLL) is employed to extend the regenerative-type fiber laser to the rational harmonic region. The frequency of the optical pulse train is split by the number of rational harmonics when the laser pulse repetition rate is greater than the driving signal of the modulator. The resulting signal is then sent back into the modulator. The pulse train has been stabilized by the application of this modified regenerative mode-locking approach. The 10 GHz electrical signal is extracted from the input 40 Gb/s optical pulse train using a base-line extraction circuit. The hybrid optical-electrical phase-locked loop serves as the foundation for this base-line extraction circuit. A temporally stable 40 Gb/s pulse train is produced at a VCO frequency that meets the fourth-order rational harmonic mode locking condition. At that instance, the base-line extraction circuit internal time delay is changed to minimize the phase detector output. The integrator output is then connected to one of the adder input ports to close the loop. A loop mirror made of several meters of polarization-maintaining fiber (PMF) can be utilized as a repetition rate doubler for high repetition rate pulse production. Researchers have previously employed the polarization-maintaining loop mirror as a notch filter to remove the initial CW signal. The phase difference between the fast and slow axes is what causes the periodic notch effect in frequency space. If the notch separation is selected to be double the low-rate pulse train mode spacing, part of the modes will be filtered out, resulting in a pulse train that passes through the loop mirror at twice the repetition rate. The notch separation becomes 80 GHz if the fiber length inside the loop mirror is around 9 m. An output at 80 Gb/s is anticipated for an input signal at 40 Gb/s. A stable dual wavelength laser has been created using the phase-locked loop (PLL), which could be significant for certain applications.

The EDFA has a maximum output of 25 dBm. Active mode locking is achieved by using a $LiNbO_3$ Mach–Zehnder intensity modulator powered by the output of a voltage-controlled oscillator (VCO) operating at 10 GHz (Alvarez et al., 2024). The VCO produces steady pulses at a precise repetition rate by means of a phase-locked loop circuit. At 20 GHz, the fiber laser is in a mode lock. In order to recombine the split pulses for mode-locked operation, the cavity length is adjusted using tunable delay lines. The cavity uses a dispersion compensating fiber (DCF) that is 70 meters long. The DCF dispersion value is -89.6 ps/nm/km. The DCF lessens gain competition in EDFA by creating a temporal difference between the pulses at two wavelengths. The stability of the pulse trains is crucial in high-speed optical transmission systems. Pulse instability in EDF-based mode-locked multiple wavelength lasers is typically caused by two primary factors. They are channel competition and cavity length drift brought on by the temperature fluctuations. The two pulses mix at the DMUX output after arriving at the MUX at various times. Even though the DCF divides the pulses at two distinct wavelengths, thermal fluctuations could cause the cavity lengths to drift. To preserve the mode locking conditions, a phase locked loop (PLL) circuit is employed in the cavity. A 10 GHz photodetector, a narrow band RF amplifier, a phase shifter, a 10 GHz mixer, a phase-locked loop controller, and a voltage-controlled oscillator (VCO) are the components of the phase-locked loop circuit. The tuning range of the VCO is 5 MHz to 7 MHz, with a central frequency of 9.953 GHz. The voltage range for tuning is 0 V to 12 V. The phase difference between the photodiode output and the VCO is detected by the mixer. The PLL controller, which consists of an analog adder circuit, a switch, and an integration circuit, receives the mixer output. One channel harmonic mode locking is achieved by carefully manipulating delay line 1, while the other channel mode locking is achieved by varying the cavity length of delay line 2 to combine the pulses. To stabilize the pulses, the PLL is then closed. Distinct MUX/DMUX channels can be used as working channels to run the laser for distinct sets of mode-locked pulse wavelengths. One-wavelength or two-wavelength operation can be achieved by varying the polarization controller (PC) in either channel because the modulator is polarization sensitive. PLL is necessary for steady pulse production. Many sidebands are observed when the PLL is absent, and since the distance between them equals the fundamental cavity frequency or the fiber laser mode spacing at 1.83 MHz, these sidebands represent the super modes of the laser. There is a suppression of the super mode noise. Fiber lasers with active mode locking are crucial for applications requiring a precise pulse rate. Among the applications are optical networking and optical time division multiplexed communication systems. There are currently commercially available mode-locked diode-pumped solid-state lasers. They offer ultra-fast, compact, and dependable lasers. Adjustments to the laser cavity are not essential, and the pulse generating process is self-starting. These lasers are being used in fiber Bragg grating-based sensors, optical communication, materials research, and tomography studies. Applications such as telecommunication, biology, materials research, and tomography all depend on short pulses. By passing through different kinds of fiber, loop mirrors, and gratings, the pulses produced by a fiber laser can be compressed.

8.8. COMPRESSION

Adiabatic soliton compression techniques, multiple-step compression based on self-phase modulation in semiconductor optical amplifiers (SOA) (Figure 8.6), highly nonlinear fibers, or noble gases, parametric gain in a fiber-optic parametric amplifier, a comb-like profiled fiber (CPF) to simulate dispersion decreasing fiber, and saturable optical absorbers, such as nonlinear optical loop mirror (NOLM) are some of the schemes.

The NOLM method can be modified to accommodate different fiber kinds, laser designs, and configurations (Zhao et al., 2021). In order to produce high-peak power pulses, gratings have been employed for pulse compression. They are also useful for extending the pulse. Many studies have been conducted on the propagation of pulses in optical fibers. Nonlinearity and dispersion effects are significant factors in pulse propagation over long fiber lengths. The dispersion selection is crucial for pulse compression. Since silica has inversion symmetry, a silica fiber second-order nonlinearity is 0. For use in ultra-long distance optical communication, soliton pulses are being thoroughly investigated. The pulses must be optically amplified at regular intervals to make up for the loss they experience because optical loss is ignored for soliton generation. Experiments on soliton transmission beyond transoceanic distances have been conducted. It has been demonstrated that error-free transmission over long distances is possible by mitigating the effects of polarization mode dispersion and soliton-soliton interaction. Erbium-doped fiber amplifiers are used in a recirculating loop to compensate for loss during the experiment. The outcome displays pulse distortion.

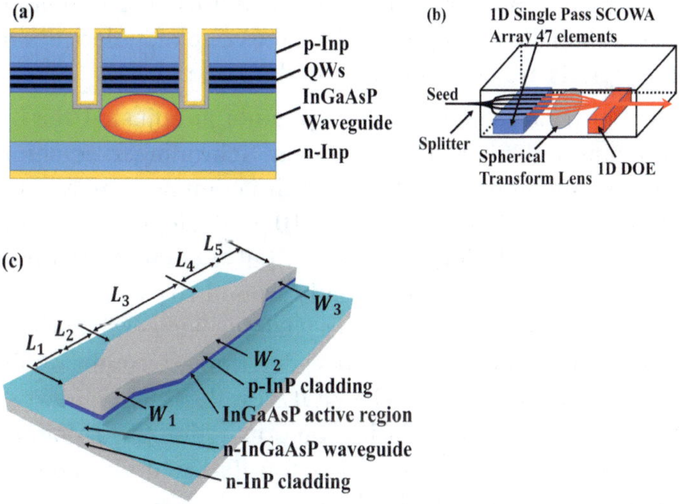

Figure 8.6. *Semiconductor optical amplifiers (a to c).*

Source: Tang, H., Yang, C., Qin, L., Liang, L., Lei, Y., Jia, P., Chen, Y., Wang, Y., Song, Y., Qiu, C., Zheng, C., Li, X., Li, D., & Wang, L. (2023). A review of high-power semiconductor optical amplifiers in the 1550 nm band. Sensors, 23(17), 7326. https://doi.org/10.3390/s23177326, Creative Commons License.

The compressed pulses are sent into a 2000 m dispersion-shifted fiber after being linearly amplified in an erbium-doped fiber amplifier (EDFA) with a maximum output of 15 dBm. This fiber dispersion slope is 0.09 ps/km/nm^2, and its wavelength for zero dispersion is 1552 nm. The dispersion-shifted fiber self-phase modulation (SPM) causes the pulses to chirp linearly while maintaining a nearly constant pulse width. To counteract the chirp and compress the pulse, a 300 m standard single mode fiber with a dispersion of 17 ps/km/nm at 1550 nm was employed. At 16 mW of input power, the pulse width is approximately 2 ps. Following the second EDFA, these 2 ps laser pulses are transmitted via a conventional single-mode fiber to accomplish high-order soliton compression. The average output power following amplification in the EDFA is 22 dBm, and there is a slight change in the pulse width. For a 30 m fiber length, the shortest pulse width of 250 fs is attained. One can compute a pulse peak power of 69 W prior to compression using the initial pulse width of 2 ps, an EDFA output power of 21.5 dBm, and a repetition rate of 1 GHz. The pulse quality is distorted by the characteristic wings linked to higher-order soliton compression. The wings are intended to be suppressed using a nonlinear optical loop mirror (NOLM). A variable attenuator, a polarization controller, a 50/50 coupler, and a 20-m dispersion-shifted fiber are part of the NOLM. The output pulses of the NOLM are wider than the input pulses since the coupler, polarization controller, and attenuator pigtails are conventional single-mode fibers.

The linear chirp produced by those pigtails is adjusted for by a 60-cm-dispersion-compensated fiber. The nonlinear Schrodinger equation is solved using the split-step method to model the entire compression system, including the NOLM. With an FWHM pulse width of 10 ps, the input pulses are transform-limited Gaussian pulses. Using a two-stage pulse compression process, gain-switched DFB laser pulses that were initially around 10 ps long were compressed into pulses with a width of about 200 fs. Using fiber nonlinearities, the pulse compression mechanism propagates across determined lengths of different fiber types and a nonlinear loop mirror. The NOLM reflectivity can exceed 95% for close to 50% and with the right polarization orientation (Avazpour et al., 2019). 25 meters of extremely nonlinear, dispersion-flattened photonic crystal fiber were utilized in the loop mirror. This kind of fiber offers low second-order and third-order dispersions, high nonlinearity, and good polarization, which are essential. Stable 40 GHz and 20 GHz RZ optical pulse trains are produced using a rational harmonic mode-locked ring laser with a Mach-Zehnder-type modulator powered by a 10 GHz RF synthesizer. To suppress self-pulsing behavior, a semiconductor optical amplifier (SOA) is employed in the cavity as a rapid saturable absorber. A comparable but less stable pulse is seen in the absence of the SOA. For the 20 GHz and 40 GHz pulses, the pulse autocorrelation trace of the pulse trains produced by the ring laser displays pulse widths of 4.4 ps and 4.1 ps, respectively. After passing through the NOLM pulse compressor, these pulses are compressed. Since the NOLM generally requires pulses shorter than two ps to obtain acceptable compression results, the initial external compression step is crucial. Following a 4 km-long dispersion-shifted fiber (DSF), where the pulse spectrum is highly expanded due to self-phase modulation, the pulse train is amplified using an EDFA to an average power of about 17 dBm. To counteract the chirp the DSF causes, a 100 m-long single-mode fiber segment comes after the DSF. It is determined that the output pulse width of the two fiber types at 20 GHz and 40 GHz is approximately 1.8 ps. After that, the pulse train is sent onto port 1 of the variable ratio coupler via an optical circulator (OC). To ensure that nearly all of the injected light emerges from port 1, the coupling ratio and the polarization orientation in the loop mirror are precisely controlled. After entering OC1 through ports 2 and 3, the reflected light enters another EDFA, which amplifies the pulse train until its average pulse energy is comparable to that of the pulse train before to entering OC1. The light enters the NOLM a second time from port 2 of the variable coupler via a second circulator (OC2). The reflected light then leaves the OC2 as an output from port 3.

8.9. MODE LOCKING

Mode-locked pulses have been produced in other ways, even though active mode locking, in which the repetition rate is dictated by the frequency of intracavity electrically driven optical modulators, makes it simpler to produce stable mode-locked pulses at a particular repetition rate. These techniques, which are commonly referred to as passive mode-locking schemes, combine highly nonlinear fibers with comb filters or saturable absorbers. When used with the active mode-locking technique, nonlinear fibers and saturable absorbers can both produce shorter pulses. It has been shown that a variety of intracavity optical filtering techniques may generate high repetition rate pulses. These consist of a micro-ring resonator, a programmable optical processor, a sampling fiber Bragg grating, and a Fabry-Perot filter. Through the filtering action, the silicon mirroring resonator or SMRR only permits the optical harmonic modes with a spacing of 2 nm to accumulate in the cavity. The FWM effect brought on by the HNLF locks the phases of these modes. The optical pulses are forced to propagate in a single direction by the polarization-independent isolator. A 1 cm segment of polarization maintaining fiber (PMF) is placed inside the cavity. A polarization-dependent grating coupler (GC) coupled to a silicon-on-insulator (SOI)-based SMRR is utilized for single-mode fiber coupling (Yeh et al., 2016). In order to create high-power pulses, relatively wide pulses are often amplified in a high-power fiber amplifier and then compressed using a grating-based pulse compressor. The sub-ps range is where the last pulse width falls. This reduces gain saturation and damage by enabling the input pulse energy distribution across a wider wavelength range to be increased. A high-power pulse amplifier and a pulse compressor are essential for producing high-power short pulses. The signal in the rare-earth-doped core co-propagates and/or counter-propagates with the pump in a conventional single-mode amplifier. The pump power that can be connected in the small core, which is normally only a few microns (3–6 mm) in diameter, limits the power output of such an amplifier. A single-mode pump laser generates the pump power. Catastrophic facet degradation typically limits the output powers of single-mode lasers. Although the output power is significantly increased by the use of many pumps, such as in a co-propagating and counter-propagating configuration, it is still only around 1 W. More pump light must be connected into the fiber for higher power applications. Fiber amplifiers provide the advantage of producing a higher-quality beam at the desired wavelength for many applications. Output levels of about 10 W might be required for free space communications, whereas kW-level pulsed or CW powers might be required for manufacturing applications. These fiber amplifiers need higher coupled pump power levels and couplings between 10 and 100 W. Coupling the pump light into the big region cladding zone is one method of increasing the coupled pump power. A multimode pump waveguide is created by encircling this huge area cladding with a second cladding zone, which is typically polymer-based. Rare earth elements like Yb, Nd, or Tm are doped into the core. The core of a single mode DCF is small enough to merely amplify the fiber basic mode. In order to generate kW-level powers, the core portions of LMA fibers usually have a diameter of 20 to 50 μm.

8.10. PUMP LIGHT

Many techniques have been investigated for connecting pump light into a large core fiber. In this kind of design, the rare earth in the core interacts with the portion of the pump power in the core to provide a signal gain. As it travels throughout the amplifier, the pump light transfers its power to the signal light. The signal output can be higher than about 1 W, which is typically possible for single-mode amplifiers. It is also possible to couple pump laser light from several semiconductor diode laser bars, each of which has an emission area of 100 μm × 0.2 μm, using a huge cladding fiber. A diode bar that is 100 μm wide generates more than 10 W of electricity (Fischer et al., 2004). With the use of appropriate pump coupling techniques, the output power of many laser diodes might be coupled into the cladding of a DCF fiber. Different pump coupling techniques and cladding-pumped high-power amplifier types have been employed. The pump moves along a multimode fiber waveguide that covers a wide area. Many pump waveguide design types have been investigated. An essential factor in pump waveguide design is improved pump coupling from diode laser bars. The basic pump waveguide has a circular form. It is composed of a large region of Ge-doped silica fiber encircled by a second cladding layer that is usually 10–20 μm thick. The rare-earth-doped core design is comparable to the single-mode DCF utilized in relatively low-power systems. Many techniques for connecting pump light into a large-area double-clad fiber (DCF) have been developed. They are typically categorized as side pumping and end pumping methods. Light from many semiconductor lasers can be coupled into the large-area due to the TFB. A tapered fiber bundle (TFB) of N fibers, each of which carries pump light, is spliced into a single output fiber to create this kind of end-pumping coupler. High coupling efficiency from all N input fibers to a single output fiber can be attained by maximizing the tapered region length. The use of a DCF fiber with a Yb-doped core in high-power fiber lasers or amplifiers has been thoroughly investigated. Investigating the relationship between ultrahigh intensity pulse lasers and matter requires the use of short pulse lasers. There are many uses for these high-intensity pulses. The creation of the ultrashort, ultrahigh-intensity pulse laser is significantly aided by the chirped pulse amplification (CPA) technology. The important component of the CPA system is the pulse compression grating. The CPA approach was initially demonstrated by Strickland and Mourou. The performance of the entire laser system is determined by the grating used to carry out the pulse compression. Especially for high-power applications, the gratings must have a high laser-induced damage threshold, a wide bandwidth, and a high diffraction efficiency.

8.11. WAVELENGTH

These wavelengths diffract at various angles when the pulse strikes the grating. The diffraction angle increases with increasing wavelength and decreases with decreasing wavelength. A pulse shorter than the incident pulse will leave the grating system. The peak intensity of the shorter exit pulse will be higher than that of the incident pulse. An incident pulse can also be converted into a longer pulse by applying the same approach. Compared to the incident pulse, the exit pulse has a lower peak power. The incident pulse passes via a pulse stretcher, which lowers the peak intensity, and an amplifier, which raises the pulse total intensity in a chirped pulse amplification (CPA) system (Strickland, 2021). Peak intensity can increase by 10–100 times using a CPA system. A chirped pulse compression (CPA) system operates on this principle. There are three different kinds of grating pulse compressors (GPCs). The popular kind of GPC utilized in the CPA system is a metal grating, typically one that is Au-coated. Over a wide bandwidth, the metal-coated gratings exhibit high diffraction efficiency. However, the application is restricted for high-power operation because of the significant metal absorption. The application of multilayer dielectric gratings is restricted to ps pulses due to their short bandwidth and low absorption. Metal multilayer dielectric grating, or hybrid grating, is a useful technique for high damage threshold and broad bandwidth. Although they are limited to low power, fiber-based pulse compressors and other nonlinear pulse compressors can also be employed. It consists of three primary components: a Pulse stretching that converts a laser low-energy, ultrashort pulse into a long pulse. Stretching reduces the peak intensity, and the long pulse duration typically ranges from 100 ps to 3 ns. A laser amplifier amplifies the stretched pulse 10–100 times, ensuring that it has enough energy. In pulse compression, an amplified long pulse with a high total energy is compressed into an ultrashort pulse that is the same length as the original pulse or less; as a result, it exhibits a short pulse and ultrahigh peak intensity. It is possible to create a laser with petawatt-level energy in picosecond pulses by using the CPA system. An ultrahigh intensity laser source is created by the technique. For many applications, such as high harmonic generation, a short pulse source with good beam quality that can be concentrated into a small spot is crucial. The guided mode of a fiber laser determines the beam quality, which is essentially power-independent. There have been reports of diffraction-limited CW power of about 10 kW when operating CW. Very high powers can be generated via a double-clad fiber design of the rod type. Because of their excellent pump coupling, rare earth-doped photonic crystal fibers, sometimes referred to as microstructure fibers, with large core and cladding diameters can generate high powers. A 200 μm inner cladding that functions as a pump waveguide surrounds the fiber air cladding, which is made up of 10 mm long, 400 nm thick silica bridges. The Yb/Al-doped core active region is made up of missing holes in the center of the fiber, spaced from corner to corner. The structure can effectively couple multimode 976 nm pump lasers due to its numerical aperture of 0.6. The small signal pump absorption is about 30 dB/m, and the fundamental mode area is 4000 mm^2. A fused silica outer covering with a diameter of 1.5 mm encloses the whole interior structure. The purpose of this outer covering is to maintain the fiber (straight) and avoid bend-induced losses.

8.12. PULSES

Ultrashort pulses with a high peak power of many gigawatts are necessary for high-field physics experiments. Fiber CPA (FCPA) is unable to meet the pulse duration criteria. To further shorten the pulses, nonlinear pulse compression techniques are required. The technique employs chirped mirror compressors for pulse compression after self-phase modulation in fibers filled with noble gases. Any spatial chirp is minimized by self-phase modulation using a waveguide. Because of their uses in spectroscopy, optical coherence tomography, and dense wavelength division multiplexing (DWDM) sources for high-capacity optical communication systems, broadband supercontinuum generation (SCG) techniques have attracted a lot of attention. For powerful narrow-band incident pulses, SC production can achieve extreme pulse spectrum broadening. Cross-phase modulation (XPM), Self-phase modulation (SPM), stimulated

Raman scattering (SRS), parametric four-wave mixing (FWM), and the fiber dispersive characteristics are thought to work together as the main mechanism of spectral broadening (Alfano and Ho, 1988). The dispersive and nonlinear characteristics of the optical fiber have an impact on the SCG operations. Supercontinuum creation requires the development of a mode-locked fiber laser with a comparatively high power. The ultrawide mid-IR supercontinuum generation in a low-loss W-type tellurite fiber that stretched from 0.9 μm to 4.2 μm and was pumped by a CNT mode locked Tm-doped fiber laser at a distance of about 2.0 μm. The transmission window of tellurite glass is roughly 5 μm. In addition to having a nonlinear index, chalcogenide materials like As_2Se_3, As_2S_3, and GaSe also have a wider transmission window, up to about 15 μm. They are crucial for the creation of supercontinuum. A flat broadened spectrum might be produced by a dispersion-decreasing fiber with a convex dispersion profile. The creation of fibers with two zero dispersion wavelengths (ZDWs) is made possible by advancements in photonic crystal fiber (PCF) design, and SCG in these fibers has been thoroughly investigated. By phase-matching over the upper ZDW, small core MOFs with two ZDWs can generate mid-IR continuum with high efficiency when utilizing appropriate pump wavelengths. Researchers discovered that the broad spectrum may be obtained with a seed wavelength that is almost midway between the ZDWs. A coherent continuum could be produced by computationally modeling a ZBLAN fiber taper with progressively decreasing core size. The nonuniform fiber taper design can gradually change the spectrum produced across the upper ZDW. Coherent supercontinuum generation in single-mode fibers employing mode-locked high-power lasers depends on nonlinear processes. These sources offer novel laser wavelengths for use in sensors, tomography, and spectroscopy. In the 1970s, semiconductor injection lasers and photodetectors were first produced for use in optical fiber transmission systems.

8.13. GAAS-BASED LASERS

GaAs-based lasers with an emission near 850 nm were utilized in old systems, whereas InGaAsP-based lasers with an emission near 1,300 and 1,550 nm were used in latest systems (Brown et al., 2021). Single wavelength lasers with distributed feedback are used in the 1,500 nm systems. These methods developed into wavelength division multiplexed (WDM) transmission, which transmits data via a number of distributed feedback lasers. The of single frequency (DFB) lasers carrying data is multiplexed into a single fiber in current systems, which use WDM-based transmission. The data rate per wavelength channel can be doubled with polarization multiplexing. The average data rate per channel is between 10 and 40 Gb/s, which adds up to a transmission capacity of about 1

Tb/s. An optical time division multiplexed (OTDM) system is an alternative to the WDM system. The data rates in an optical time division multiplexing (OTDM) system are high enough to enable optical multiplexing, which creates high data rate signals from multiple low data rate signals, and demultiplexing, which converts high data rate signals into multiple low data rate signals. An optical clock, or timing signal, must be taken out of the incoming high speed optical data in order to perform the demultiplexing operation. Cross-phase modulation, cross-gain modulation, and four-wave mixing have been used in this clock recovery procedure. High data rate signals are created during the optical time division multiplexing process. The procedure uses a mode locked laser to produce short pulses at a comparatively high repetition rate. The information is encoded using a synchronized 20 Gb/s optical modulator and sent over several delay lines with appropriate values to produce a high data rate signal. A mode-locked Er-doped fiber laser with an intracavity $LiNbO_3$ modulator is used to generate the initial pulses. The ring cavity produces pulses at various wavelengths using a tunable filter. The emission of the pump laser is close to 1,480 nm. All passive components are used by the parallel-type multiplexer. To manage the relative time delay between each channel, a stabilization strategy is necessary, which makes the configuration difficult. A modulated channel is created in the high-bit-rate signal by combining a high-speed optical clock pulse train with modulated signal pulses, which are then introduced to an optical switch. All of the channels can be individually modulated by cascading this procedure. At the SOA output, a four-wave mixing (FWM) signal is generated by the input optical signal and the local oscillator optical signal. In a transmission system, the input optical data is demultiplexed to lower data rate channels using the timing signal (clock), which is then monitored using conventional electronic circuits to determine the error rate. An optical phase-locked loop combined with an electrical low-frequency phase-locked loop could be used to perform the clock recovery procedure. Four-wave mixing in a semiconductor optical amplifier (SOA) can be used to do optical demultiplexing, which is the process of creating multiple lower data rate signals from a high data rate signal. Four-wave mixing occurs in optical fibers as well as semiconductor optical amplifiers. Long fiber lengths are required for four-wave mixing in fibers.

8.14. SEMICONDUCTOR OPTICAL AMPLIFIER (SOA)

Using a SOA, this method enables the simultaneous extraction of any regularly spaced collection of bits from the incoming data stream (Connelly, 2002). A SOA-based clock recovery procedure uses four-wave mixing (FWM) to extract the clock from the sent data. The optical clock is provided by a mode-locked fiber laser running at 10 GHz, the precise frequency of which is determined by phase locking to the 400 Gb/s data that was received. Several research labs are currently working on developing very fast OTDM transmission systems. The creation of these systems heavily relies on semiconductor optical amplifiers. They can be applied to optical clock recovery as well as optical demultiplexing. Ultra-large capacity transmission networks, or those with capabilities close to Tb/s, are becoming more in demand. Since it can lower the number of WDM channels, high-speed transmission with bit rates of multiples of 10 Gb/s is highly attractive. For these systems, transmission via optical solitons and time-division multiplexing are crucial. In dispersion-managed systems, soliton transmission enables ultra-long-distance transmission without errors. To keep the overall system dispersion near zero, this technique uses a dispersion compensation fiber (DCF) with negative dispersion to periodically counteract the cumulative chromatic dispersion. With nearly zero dispersion, this system also reduces the Gordon-Haus timing jitter. A recirculating loop experiment was conducted to replicate transmission over long distances. The optical data in a loop experiment repeats itself many times. The loop contains all the components needed for a transmission experiment, including fibers and an amplifier chain. The error rate test device, the transmitter, and the receiver are not included in the loop. To load the loop with data, the transmitter switch is turned on at the beginning of the experiment. For a predetermined number of revolutions, the data is permitted to move around in the loop. The overall transmission distance is equal to the number of revolutions. Loop length provides information on the delay per loop circulation. A switch releases the data from the loop after a predetermined total transmission distance, and the error rate is then calculated. In the soliton experiment, the recirculating loop is made up of many units. A dispersion-shifted fiber (DSF), a DCF, and a 980 nm laser-pumped erbium-doped amplifier (EDFA) are included in each unit. It also features an optical band-pass filter (OBPF) and a polarization controller. The transmitter uses optical time division multiplexing (OTDM) of 10 Gb/s RZ (return to zero) data pulses to create a 20 Gb/s optical soliton data stream. A distributed feedback (DFB) laser, sinusoidally driven electroabsorption (EA) modulators, and two $LiNbO_3$-intensity modulators running at 10 Gb/s with a pseudorandom binary sequence (PRBS) input are the sources of these data pulses. The two 10 Gb/s PRBS signals are multiplexed to produce the 20 Gb/s data. The optical 20GHz PRBS signal can be generated using a 20GHz $LiNbO_3$ modulator or electroabsorption modulator powered by a 20GHz electrical PRBS data after a 20GHz pulse from a stabilized mode-locked fiber laser. About 9 ps is the generated pulse width. These pulses go via a 20 GHz clock-driven $LiNbO_3$ phase modulator. An electric phase shifter is used to modify the delay between the optical data bits and the phase modulation signal in order to optimize the phase condition. The transmission performance is effectively enhanced by this first phase modification. The optical time division multiplexing (OTDM) of 20 Gb/s transmissions produces a 40 Gb/s signal. The clock signal is extracted from the transmitted signals in the receiver.

CHAPTER 9

ADVANCES IN SEMICONDUCTOR LASERS

Contents

9.1. INTRODUCTION

Broad-area, high-power semiconductor lasers are desirable sources for laser pumping, aerospace, and material processing. Since the temperature has a major impact on the quantum well performance, including its internal quantum efficiency and mode gain, the active region design is essential to achieving the necessary high power and electro-optical conversion efficiency. Some researchers investigated the theoretical and experimental effects of temperature on the active region of a high-power semiconductor laser operating at 808 nm. Quasi-3D models, which integrated novel mathematical techniques with full steady-state semiconductor and carrier confinement efficiency, were used to conduct the simulations (Alonso-Álvarez et al., 2018). In the temperature range of 213 to 333 K, the critical aluminum content of the quantum barrier was suggested, and the relationship between temperature and different loss sources was presented. This information offers directions for designing the epitaxial structures of high-power semiconductor lasers under various operating conditions.

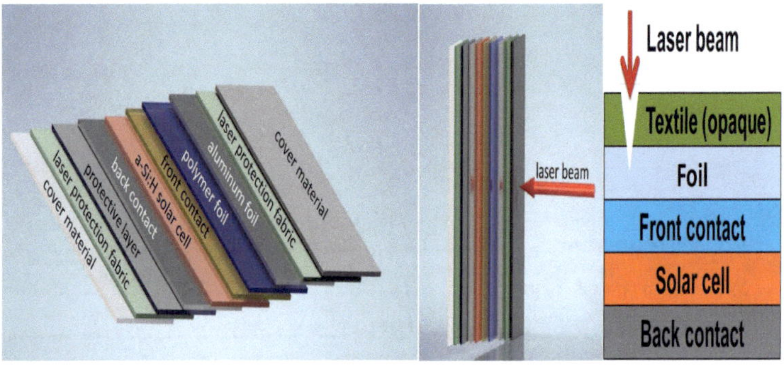

Figure 9.1. *Laser bar chips.*

Source: Gawlik, A., Brückner, U., Schmidl, G., Wagner, V., Paa, W., & Plentz, J. (2023). Amorphous silicon thin-film solar cells on fabrics as large-scale detectors for textile personal protective equipment in active laser safety. Materials, 16(13), 4841. https://doi.org/10.3390/ma16134841, Creative Commons License.

9.2. LASER BAR CHIPS

Standard laser bar chips (Figure 9.1) with a cavity length of 2 mm were then fabricated using the optimized epitaxial structure. At an injection current of 400 A, high electro-optical conversion efficiency was achieved, while the carrier confinement efficiency reached up to 98%. It is difficult to design the active region at a certain temperature since there is no clear explanation of how temperature and barrier height affect carrier leakage loss or how the active region gain varies with temperature and materials. Because of its exceptional Raman and thermal properties, chemical vapor deposition (CVD) diamond crystal is regarded as the perfect material platform for Raman lasers with high power and good beam quality.

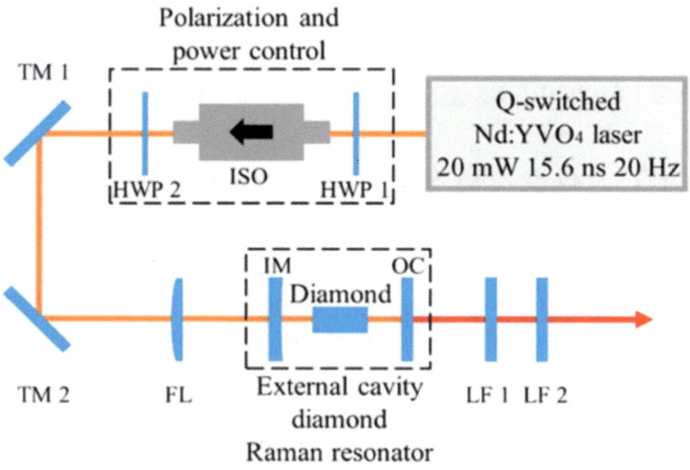

Figure 9.2. *Diamond Raman lasers.*

Source: Liu, Y., You, W., Zhu, C., Li, M., Sun, Y., Yin, X., Chen, D., Feng, Y., Chen, W., & Yang, X. (2022). A review of ns-pulsed Raman lasers based on diamond crystal. Frontiers in Physics, 10. https://doi.org/10.3389/fphy.2022.1054234, Creative Commons License.

As CVD diamond crystal formation technology continues to advance, diamond Raman lasers (DRLs) (Figure 9.2) have demonstrated benefits in terms of wavelength expansion, high beam quality, and power operation. However, the detrimental impact of the thermal impact on the beam quality is getting higher as the output power of DRLs approaches the kilowatt level. Scientists developed a thermal-structural coupling model with the goal of articulating the fundamental limitations of the thermal effects for high-power DRLs. The influence of pump power, cavity structure, and crystal size have been methodically examined using this model. An asymmetrical concentric cavity has a high thermal influence on the device than a symmetrical concentric cavity. The transient simulation also demonstrates that the pump power has no effect on the heating and cooling of DRLs, with steady state times of just 1.5 and 2.5 ms, respectively. It was discovered that reducing the thickness of the crystal and widening the crystal, or raising the curvature radius of the cavity mirror help to lessen the device thermal impact. Many industries, including laser medicine, eye-safe lidar, laser satellite communications, space exploration, etc., have used high-power lasers with good beam quality. Raman lasers have drawn a lot of interest in the development of high-beam quality and high-power operating lasers due to their ability to purify beams through the employment of special automatic phase-matching properties. Because of its high Raman gain coefficient, low Raman linewidth, low thermal expansion coefficient, and thermal conductivity, which is higher than that of other crystal materials, diamond crystal is regarded as the perfect material for high-power and good beam quality Raman lasers.

9.3. DIAMOND RAMAN LASERS (DRLS)

DRLs with high power and good beam quality have advanced significantly due to advancements in the chemical vapor deposition (CVD) fabrication (Figure 9.3) process of large-size single-crystal diamonds (Arnault et al., 2022). A first-order Stokes light output with a power of 108 W was reported in 2014. DRL output power has exceeded 100 watts, with a beam quality factor of M2 < 1.1 and an optical-optical conversion efficiency of 34%. With a beam quality factor of M2 = 1.1 and an optical-optical conversion efficiency of 36%, researchers achieved a second-order Stokes light output of 302 W in 2018.

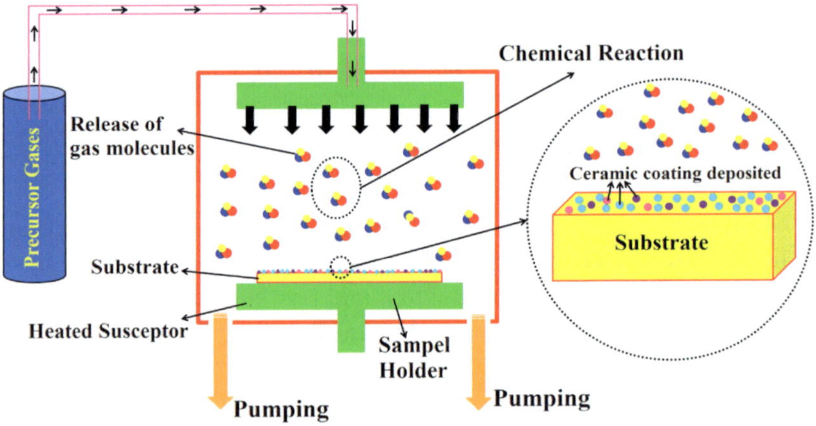

Figure 9.3. *Chemical vapor deposition.*

Source: Sabzi, M., Mousavi Anijdan, S. H., Shamsodin, M., Farzam, M., Hojjati-Najafabadi, A., Feng, P., Park, N., & Lee, U. (2023). A review on sustainable manufacturing of ceramic-based thin films by chemical vapor deposition (CVD): Reactions kinetics and the deposition mechanisms. Coatings, 13(1), 188. https://doi.org/10.3390/coatings13010188, Creative Commons License.

Despite the developments, it is evident that the laser beam quality starts to decline as DRL output power approaches the kilowatt level, suggesting that the negative impact of thermal effects on beam quality is growing. The primary cause of the thermal effects is a temperature increase brought on by the crystal conversion of some of the energy used for the pump light into heat. A decrease in the conversion efficiency of the pumped light within the laser crystal will arise from the uneven temperature distribution inside the crystal, which will also produce uneven expansion within the medium and create thermal stress to deform the crystal. The size of the laser crystal, the power and beam distribution of the pump light, the laser resonant cavity configuration, the heat dissipation method, and the structure of the heat dissipation system are significant factors influencing the thermal effect of the laser crystal. Due to diamond exceptional thermal conductivity, which permits its thermal effects to be ignored at low input power, less research has been done on the thermal impacts of DRLs. However, when the output power of DRLs reaches the kilowatt level, the thermal effects of the diamond play a crucial role in restricting the laser output power and beam quality.

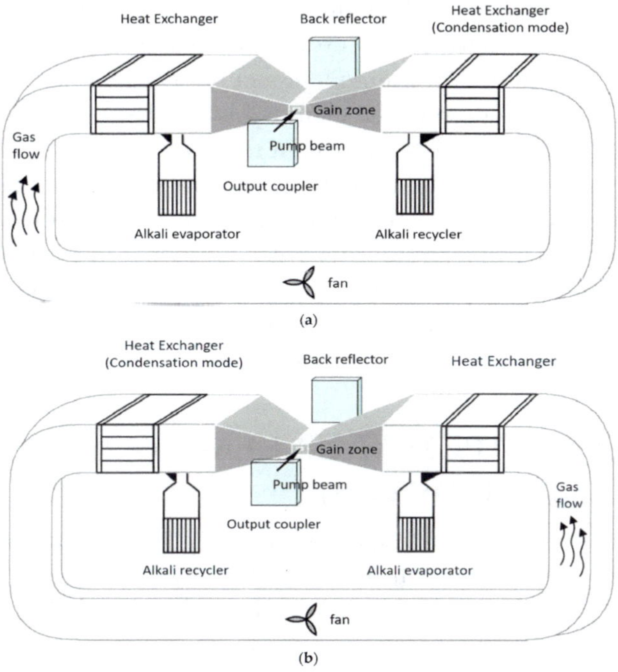

Figure 9.4. *Diode pumped alkali metal vapor lasers (a and b).*

Source: Liu, S., Tan, R., Xu, W., Ning, F., & Li, Z. (2024). Double-cycle alternating-flow diode pumped potassium vapor laser. Photonics, 11(5), 391. https://doi.org/10.3390/photonics11050391, Creative Commons License.

Researchers investigated the thermal effects of DRLs in 2021 using a simpler point source as a heat source model and discovered that the device temperature rose to 70 K at an output power of 302 W. They used the heat source model with crystal internal heat transfer and discovered that the device temperature rose to just 7.15 K at an output power of 132 W. We can establish a thermal-structural coupling model to know the limits of the thermal effects for high-power devices. Based on this model, scientists examined the relationship between device thermal effect and pump power, and also studied the influence of cavity structure, including cavity type, diamond size, and cavity mirror radius of curvature. The maximum temperature inside the diamond crystal varies with length. As the crystal length extends, it is evident that the maximum temperature inside the crystal steadily drops from 310.1 K to 306.3 K. This is due to an increase in crystal length that will improve the conversion of pump light to Stokes light, hence increasing pump light usage and decreasing heat accumulation. It is evident that the maximal thermal stresses within the crystal are not much affected by crystal length. As the crystal length rises, the ultimate thermal deformation progressively drops. When the width of the crystal is fixed, it is evident that the maximum temperature rises with increasing crystal thickness because the bottom of the diamond is fixed at 298 K, and increasing the thickness of the diamond is equivalent to increasing the distance between the heat source and the heat sink. This suggests that the thermal effect of DRLs increases with increasing crystal thickness. When the width of the crystal is fixed, the maximum temperature falls with increasing crystal width because expanding the width is equivalent to increasing the contact area between the diamond and the heat sink. It is evident that as crystal width increases, the device maximum thermal stress slightly reduces.

9.4. DIODE PUMPED ALKALI METAL VAPOR LASERS (DPALS)

The benefits of DPALs (Figure 9.4) have garnered a lot of interest in recent years (Pitz and Anderson, 2017). Applications for DPALs have been demonstrated to exist in the aerospace, military, medical treatment, and industrial processing and several experiments employ a variety of gain media types, and each material has a unique absorption wavelength. Cesium (Cs) requires an absorption wavelength of around 852 nm, rubidium (Rb) requires a wavelength of 780 nm, and potassium (K) requires a wavelength of 766 nm. Quantum efficiencies are higher than those of conventional solid state or fiber lasers, with rates of 95.27% for Cs, 98.14% for Rb, and 99.47% for K compared to 75.94% for Nd:YAG. In very high-power laser systems, low quantum defect plays a key role in lowering the thermal effect and increasing overall laser efficiency. High-power DPAL is still being developed. A common high-power diode laser array may reach a spectral linewidth of roughly 3 nm (FWHM) in the near-IR spectrum, which is one important factor. An inconsistency between the pump and absorption spectra results from DPAL narrow absorption spectrum. The smile effect of a diode laser array is another component that influences the absorption efficiency. The diode laser array spectral linewidth is reduced using volume Bragg grating (VBG) and external cavity feedback technology. Arrays of diode lasers with a lower smiling effect are selected. Researchers have investigated the high-power narrow-linewidth diode laser for alkali metal vapor laser pumping. They used an external cavity with a holographic grating to demonstrate laser diode array line narrowing. By adding a plane reflective mirror to an exterior cavity in a Littrow configuration to improve the correlation between emitters, researchers were able to minimize the smile effect in spectral linewidth narrowing on high-power laser diode arrays. A high-power diode laser with a Faraday anomalous dispersion optical filter that is wavelength-locked and spectrally constricted was described by Hao Tang. The common diode laser wavelengths of 852 and 780 nm serve as the foundation for most DPAL research. The advancement of material growth and device process technologies has led to the commercialization of 766 nm diode lasers. The optical correction approach is a good method to lessen the smile effect. The design needs a lot of optical lenses, which increases its complexity and makes it harder to modify.

9.5. APPLICATIONS

For mobile electronic devices to be charged safely and effectively, wireless charging technologies are essential. Present wireless charging methods that involve microwave transmission, magnetic resonance coupling, and inductive coupling are heavy, necessitate complex systems, expose users to radiation, and have relatively limited energy transfer ranges. By refining the external cavity structure of semiconductor lasers for laser charging applications, researchers devised a long-distance optical power transfer system. A 10 m transmission distance is intended for an ultra-long stable oscillating laser cavity. An enhanced laser cavity that generates an output of 2.589 W at a transmission distance of 150 cm is created by simulating the structural factors for stable operation in order to identify the ideal laser cavity design. The ability to achieve ultra-long-distance power transmission using this wireless power transfer system based on a vertical external cavity surface-emitting laser is demonstrated by the peak power attenuation, which is only around 6.4% as the transmission distance grows from 50 to 150 cm. The VECSEL laser-based wireless energy transmission system is the first of its kind to yield an output power of more than 2.5 W over a 1.5 m transmission distance (Michalzik, 2013). This portable, and secure long-distance wireless laser energy transmission technology is better suited for interior charging applications than previous wireless energy transmission methods. Automated trains, mobile intelligent devices, railway transportation, and other innovations that greatly improve people lives have been made possible by the quick growth of the 5G network and Internet of Things technology. User quality of life or QOLs have also been improved by the growing diversity of communication and entertainment systems brought about by the use of billions of mobile devices. However, carrying a charger that must be physically attached to a power outlet might be hard because mobile devices must be charged on a regular basis. Electronic devices can receive electrical energy from a power source via wireless power transfer technology without any physical contact. Studies on mobile device wireless charging technologies have grown recently. Magnetic resonance coupling, inductive coupling, and microwave radiation are the three main forms of wireless charging systems. The charging distance is quite low, usually within a few centimeters, but inductive coupling is safe and requires only basic equipment. Effective energy transmission can be achieved using magnetic resonance coupling, although problems like a large coil volume and a short charging distance arise. Microwave radiation systems employ microwaves as the energy transmission channel instead of a variable or fluctuating magnetic field. Although microwave radiation systems have a low energy transmission efficiency, they can carry energy over thousands of meters. Microwave radiation is only appropriate for industrial applications because to dense high radio frequency leakage that occurs during transmission and is damaging to human health. Through collimation, laser beams with high directivity can be used to transmit optical power over great distances. Energy transfer methods that use light as the energy carrier are strongly supported by the development of laser technology. Laser distinct qualities make them appropriate for transmitting optical power wirelessly.

9.6. TRANSMISSION

Long-distance energy transmission is made possible by lasers directivity, which enables them to have a beam diameter of a few centimeters at a transmission distance of several kilometers. Solar cells with the appropriate wavelengths at the receiving end may effectively absorb and transform the energy of monochromatic and high-power density laser beams into electrical energy. The resonant cavity structure of lasers reduces the possibility of laser leakage. The laser oscillation and output stop when a barrier of any kind interferes with the laser resonator. The safety of laser energy transmission systems can be increased by employing a resonant cavity as an energy transmission channel. Because of their high power, affordability, and adaptable external cavity structures, optically pumped vertical external cavity surface-emitting lasers (VECSELs) are the perfect light source for secure and effective wireless optical power transmission systems. VECSELs offer high output power and superior beam quality in a portable configuration by combining the benefits of gas and solid-state lasers (Rahimi-Iman, 2016). These distinct external cavity design enables the insertion of optical components for functions such as mode locking and frequency conversion. Light can be employed as an energy and information carrier to transport power and information at the same time by incorporating an electrooptic modulator into the cavity. VECSELs can produce emissions from ultraviolet to near-infrared wavelengths by fusing intracavity frequency conversion systems with semiconductor band engineering designs. VECSELs external cavities can be made to fit the requirements of different applications. The mid-infrared band ECQCLs have many distinct spectral characteristics that are useful for diverse applications, including spectroscopy, protein detection, medical diagnosis, free space optical communication, and others.

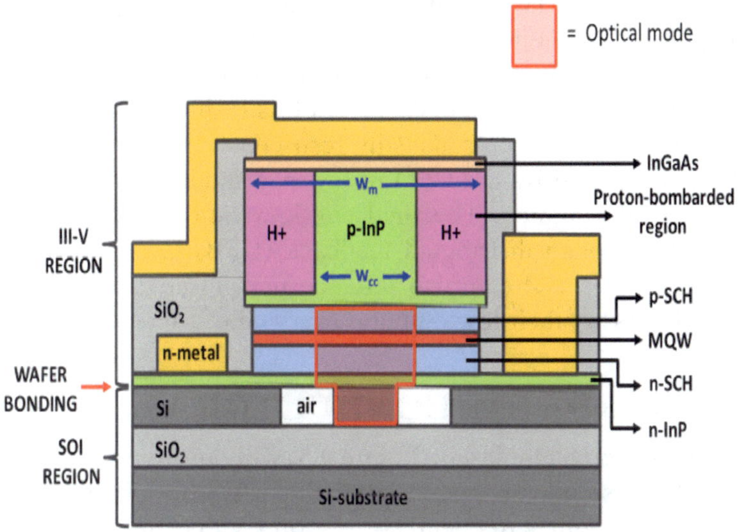

Figure 9.5. *Infrared semiconductor laser.*

Source: Buffolo, M., De Santi, C., Norman, J., Shang, C., Bowers, J. E., Meneghesso, G., Zanoni, E., & Meneghini, M. (2021). A review of the reliability of integrated IR laser diodes for silicon photonics. Electronics, 10(22), 2734. https://doi.org/10.3390/electronics10222734, Creative Commons License.

The acousto-optic modulator (AOM) control structure, multimode interference (MMI) structure, and intra-cavity out-coupling structure are a few of the novel structures. The diffraction gratings can be used as optical feedback elements to accomplish wide tuning and at a low cost, both of which help to simplify the laser structure. The common approach to ECQCLs is still the employment of diffraction gratings as optical feedback elements, and ECQCLs provide an additional choice for multi-component detection. III-V PN junction semiconductor lasers based on quantum wells typically have an operational wavelength of no more than 4 μm, which is insufficient to meet mid-far-infrared wavelength requirements. The high-quality operation of the laser cannot be ensured due to defects, significant Auger recombination loss, and free carrier absorption loss. The nonlinear optical processes that result from the nonlinear polarizability of materials can be used to create a mid-infrared laser with a broad wavelength. One common use of difference frequency generation (DFG) is the optical parametric oscillator (OPO), which combines a visible and a near-infrared laser to produce a mid-infrared laser with the frequency. OPO system strong accessibility to a wide range of pulse forms is its main advantage. A powerful peak power of several MW in pulse mode can be produced by employing an ultra-short pulse laser on a femtosecond scale. These characteristics are very beneficial for applications using nonlinear microscopy. The complexity of the optical system in OPO systems makes them less resilient, and the wavelength accessibility of nonlinear crystals still restricts the tuning range. A pump laser system is required by the OPO in order to induce the nonlinear optical process by optical pumping. The final configuration is huge and unsuitable for portability. Although distributed feedback (DFB) lasers are excellent for spectroscopic sources and can also produce mid-infrared wavelengths, their primary drawbacks are their low output power and restricted tuning ability.

9.7. INFRARED SEMICONDUCTOR LASERS

Mid-infrared semiconductor lasers (Figure 9.5) without any additional lasers are typically small and simple to operate, but their performance is constrained since the emission wavelength is frequently a spectrum with many laser modes rather than a single laser mode. A type of Fabry-Perot (FP) laser is inappropriate for use in spectroscopy. An optical gain medium, an optical device for linking the output of the gain dielectric waveguide to the external cavity free space mode, and a wavelength-selective element like a diffraction grating or interference filter make up an ECQCL, or wavelength-tunable laser. It is also possible to incorporate other optics like prisms, beam splitters, and polarizers. When compared to FP lasers, ECQCL significantly increases the laser linewidth in addition to improving tunability (Wysocki et al., 2008). A laser external cavity and optical feedback components like collimating lenses and wavelength selectors often make up the fundamental construction of ECQCLs. The laser is improved by the feedback of the wavelength selectors, which can be diffraction gratings, photoelectric filters, acousto-optic filters, etc. By changing the grating angle, the spectrum can be reflected, allowing the laser output wavelength to be changed. A unipolar semiconductor device called QCL emits light by using electron transitions between sub-bands that are restricted in the heterostructure conduction band. By varying the thicknesses of quantum wells and barriers, a QCL electron distribution must be engineered to have a population inversion. The sequential wavefunctions, which indicate the electron probability, are followed by injected electrons as they move from higher levels to lower levels. The electrons optically transit at the active region during this procedure. Light emission can be increased by sequentially piling the active section of the similar design. Laser oscillation is produced when light from one active zone promotes subsequent photon radiation. While producing the stimulated emission of photons and electrons, the electrons move from one active region to the next during this process. The laser device is known as a quantum cascade laser. The number of electrons that can contribute to the stream for light emission increases with the number of cascades constructed. Higher quantum efficiency in laser operation is provided by this optical amplification method.

By using a completely alternative method for laser emission that is not dependent on the band gap, QCLs circumvent the working principle of traditional semiconductor lasers.

9.8. QCLS

By forming the active region into nanometer-thick quantum wells, these energy levels are produced artificially. The thickness of the wells and the height of the energy barrier separating them determine the difference in the energy levels that are used to quantify and characterize the electron mobility perpendicular to the layer interface. Utilizing optical transitions between quantized electronic states to decouple lasing emissions from the bandgap, QCLs function similarly to a laser but have considerably better features and operational characteristics compared to semiconductor lasers. A mechanically controlled grating-based ECQCL primary drawback is its comparatively slow wavelength tuning, which is usually in milliseconds.

Fast tuning across a wide mid-infrared spectrum is necessary for various infrared defense applications. Using an electrically controlled AOM allows for faster wavelength tuning than traditional ECQCLs with mechanically controlled gratings. An electronic signal is applied to a piezoelectric transducer that is coupled to an optical crystal, like germanium, that is transparent at the wavelengths required for the laser to function in an AOM to produce radio-frequency acoustic waves. The phase transmission grating from which the light beam traveling through the AOM crystal can be diffracted is represented by the acoustic wave.

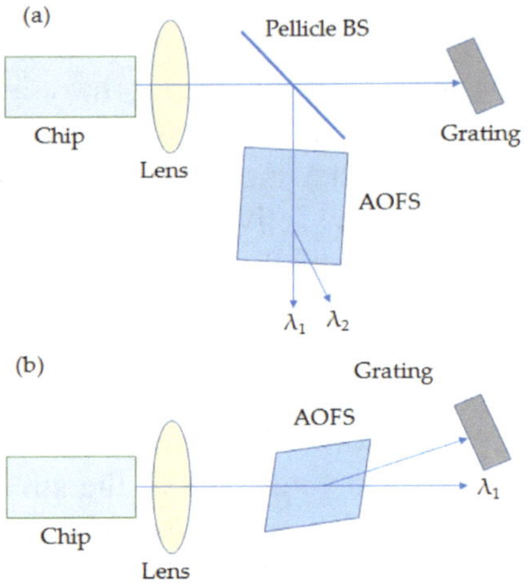

Figure 9.6. *ECQCL spectroscopy (a and b).*

Source: Ma, Y., Ding, K., Wei, L., Li, X., Shi, J., Li, Z., Qu, Y., Li, L., Qiao, Z., Liu, G., Zeng, L., & Xu, D. (2022). Research on mid-infrared external cavity quantum cascade lasers and applications. Crystals, 12(11), 1564. https://doi.org/10.3390/cryst12111564, Creative Commons License.

Higher conversion efficiency, small size, and reliability are benefits of ECQCLs. High sensitivity and selectivity are two benefits of using adjustable ECQCL spectroscopy (Figure 9.6) (Akhgar et al., 2020).

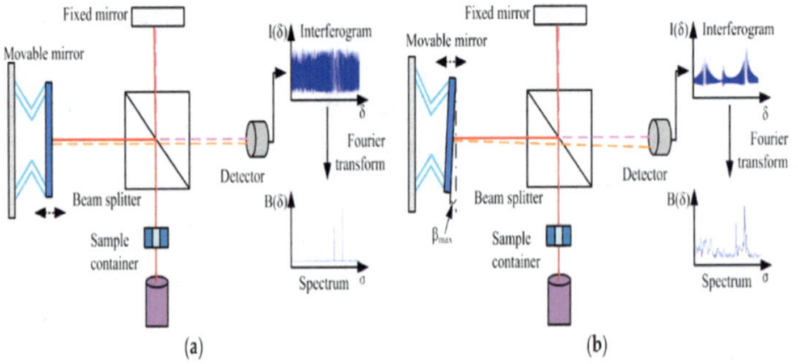

Figure 9.7. *Fourier transform infrared spectroscopy (a and b).*

Source: Fadlelmoula, A., Pinho, D., Carvalho, V. H., Catarino, S. O., & Minas, G. (2022). Fourier transform infrared (FTIR) Spectroscopy to analyze human blood over the last 20 years: A review towards lab-on-a-chip devices. Micromachines, 13(2), 187. https://doi.org/10.3390/mi13020187, Creative Commons License.

It does not require sample preparation, is quick, and is non-destructive. The use of ECQCLs for gas detection by spectroscopic groups has grown popular due to advancements in ECQCLs in terms of tunability, output power, reliability, and operating temperature. ECQCLs were used to demonstrate several absorption spectroscopy techniques. Every substance has distinct functional groups, and for different wavelength lasers, different functional groups have distinct optical activity and absorption properties. The robust detection of analytes in strongly absorbing matrices, like water, is made possible by the longer relevant path length when compared to the conventional standard Fourier transform infrared spectroscopy (FTIR) (Figure 9.7).

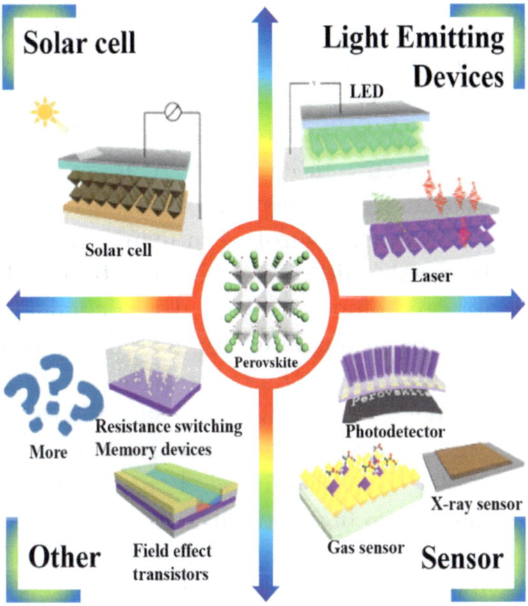

Figure 9.8. *Perovskite laser.*

Source: Cheng, M., Jiang, J., Yan, C., Lin, Y., Mortazavi, M., Kaul, A. B., & Jiang, Q. (2024). Progress and application of halide perovskite materials for solar cells and light emitting devices. Nanomaterials, 14(5), 391. https://doi.org/10.3390/nano14050391, Creative Commons License.

The inherent drawbacks of laser-based infrared spectroscopy have been addressed by recent developments in laser technology, which have also produced a broad range of accessible spectral coverage. ECQCLs have a smaller spectral coverage than FTIR due to their narrow emission wavelength range, but they have the important benefit of being stable and practical for amide detection. The infrared absorption spectrum of water-based solutions, including bodily fluids like blood, serum, and breast milk, as well as foods like milk, can be directly measured using the large optical path. The increased spectrum coverage, which includes the protein infrared band, offers benefits for both qualitative and quantitative protein research when compared to previous laser-based infrared lasers. ECQCLs will be employed in the future to investigate stoichiometry-based protein quantification in complex matrices and dynamic secondary structure modifications.

9.9. PEROVSKITE LASERS

A wide range of perovskite lasers (Figure 9.8) have surfaced in recent years, including variations in preparation methods, pump source wavelength and pulse width, and resonator type. However, the performance of the laser itself and the practical use of perovskite lasers are still in their early phases. Solution-processed semiconductors have the advantage of having adjustable optoelectronic characteristics. The optical band gaps and energy levels of semiconductor materials can be controlled in many ways. Organic materials, inorganic nanocrystals, metal-halide perovskites (MHPs), and quantum dots are included in the category of solution-processed semiconductors; each of these materials exhibits optoelectronic tunability. Compared to conventional crystalline inorganic semiconductor materials, solution-processed semiconductor materials as optical gain media provide numerous advantages. There are several types of solution-processed semiconductors that can achieve light emission of the entire spectrum from ultraviolet to near-infrared. Large absorption and radiation cross-sections are conducive to high optical gain. Metal halide perovskites are less stable despite having superior lasing capabilities. Light, heat, oxygen, and water can break down metal halide perovskite. Water reacts with $MAPbI_3$ and breaks it down to create PbI_2 and MAI, where the MAI yields hydrogen iodide and volatile methylamine (Shallcross et al., 2019). Through iodide vacancies, oxygen enters the perovskite, trapping electrons to create the extremely reactive superoxide anion. $MAPbI_3$ is broken down by the superoxide anion into PbI_2, I, and H_2O. The breakdown of $MAPbI_3$ is significantly accelerated when oxygen and light are present at the same time. After chemical breakdown, MAI and HI are sublimated to complete the thermal breakdown of $MAPbI_3$. Researchers have created a number of techniques to increase the stability of metal halide perovskite lasers. Device encapsulation can shield perovskites from oxygen and moisture exposure, halting the irreversible loss of volatile species due to heat and light. Perovskite lasers have high stability when shielded by DBR cavities, boron nitride films, and polymers. Reducing the thermal degradation of perovskite by boosting heat dissipation will increase the stability of the perovskite laser. Heat dissipation can be accelerated by covering a perovskite laser in a boron nitride layer with strong thermal conductivity. One efficient way to achieve laser stability is to synthesize perovskite single crystals with low trap density, which can prevent oxygen and light from degrading the perovskite structure. Research on hybrid and inorganic perovskite semiconductors has demonstrated that room temperature optically pumped lasering emission is possible over the visible spectrum. Perovskite lasers are often separated into low-dimensional single crystal and polycrystalline thin-film types based on the various crystal types (Zhong et al., 2020).

The hybrid perovskite material provides high intensity, high stability, and easy control. Single-mode perovskite lasers based on polycrystalline thin films have also been the subject of many studies in recent years, in addition to the multi-mode lasers based on mirror microcavities. Low-dimensional single-crystal perovskite nanomaterials can use their own regular form and smooth interface to create a good optical resonator and effectively manage incident light through the excitation resonance effect, in contrast to polycrystalline thin-film perovskite materials. The low threshold value and small volume are benefits of the laser based on a low dimensional single crystal perovskite nanostructure. One-dimensional semiconductor nanowires are small and have the ability to spontaneously form a Fabry-Perot cavity and be used as gain material. In order to achieve laser emission, the light wave can be reflected back and forth between the two ends of the nanowires, continuously gaining and forming a steady wave with the same frequency and phase. The luminescence behavior of nanowires gradually shifts from spontaneous radiation to stimulated amplification radiation as pumping energy increases, and the luminescence peak line width narrows. There are still many problems to be resolved in spite of the significant advancements in perovskite semiconductor materials and perovskite semiconductor lasers. The produced perovskite single crystal samples exhibit some degree of unpredictability, regardless of whether they were created via solution crystallization or chemical vapor deposition. We should investigate a technique for producing uniform perovskite samples in order to achieve the large-scale and commercialization of perovskite lasers.

BIBLIOGRAPHY

1. Adams, A. R., O'Reilly, E. P., & Silver, M. (1999). Chapter 2—Strained layer quantum well lasers. In E. Kapon (Ed.), Semiconductor Lasers I (pp. 123–176). Academic Press. https://doi.org/10.1016/B978-012397630-7/50002-0 (accessed on 30th July 2025).
2. Agrawal, G. P., & Dutta, N. K. (1993). Laser structures and their performance. In G. P. Agrawal & N. K. Dutta (Eds.), Semiconductor Lasers (pp. 180–230). Springer US. https://doi.org/10.1007/978-1-4613-0481-4_5 (accessed on 30th July 2025).
3. Ahmed, M. F., Bakry, A. H., & Albelady, F. T. (2014). Digital modulation characteristics of high-speed semiconductor laser for use in optical communication systems. Arabian Journal for Science and Engineering, 39(7), 5745–5752. https://doi.org/10.1007/s13369-014-1120-9 (accessed on 30th July 2025).
4. Akasaki, I. (2007). Key inventions in the history of nitride-based blue LED and LD. Journal of Crystal Growth, 300(1), 2–10. https://doi.org/10.1016/j.jcrysgro.2006.10.259 (accessed on 30th July 2025).
5. Akhgar, C. K., Ramer, G., Żbik, M., Trajnerowicz, A., Pawluczyk, J., Schwaighofer, A., & Lendl, B. (2020). The next generation of IR spectroscopy: EC-QCL-based mid-IR transmission spectroscopy of proteins with balanced detection. Analytical Chemistry, 92(14), 9901–9907. https://doi.org/10.1021/acs.analchem.0c01406 (accessed on 30th July 2025).
6. Alfano, R. R., & Ho, P. P. (1988). Self-, cross-, and induced-phase modulations of ultrashort laser pulse propagation. IEEE Journal of Quantum Electronics, 24(2), 351–364. https://doi.org/10.1109/3.132 (accessed on 30th July 2025).

7. Alonso-Álvarez, D., Wilson, T., Pearce, P., Führer, M., Farrell, D., & Ekins-Daukes, N. (2018). Solcore: A multi-scale, Python-based library for modelling solar cells and semiconductor materials. Journal of Computational Electronics, 17(3), 1099–1123. https://doi.org/10.1007/s10825-018-1171-3 (accessed on 30th July 2025).

8. AlSaffar, D., & Jihad, N. J. (2024). Design analysis for wave length division multiplexing technique in optical communications systems. Journal of Optics. https://doi.org/10.1007/s12596-024-02319-6 (accessed on 30th July 2025).

9. Alvarez, J., Leyva, W., Racedo, F., & Sarmiento, R. (2024). Development of a compact control circuit for voltage-controlled oscillator with applications in laser stabilization. AIP Advances, 14(8), 085303. https://doi.org/10.1063/5.0217529 (accessed on 30th July 2025).

10. Amann, M. C., & Ortsiefer, M. (2006). Long-wavelength ($\lambda \geq 1.3$ μm) InGaAlAs–InP vertical-cavity surface-emitting lasers for applications in optical communication and sensing. Physica Status Solidi (a), 203(14), 3538–3544. https://doi.org/10.1002/pssa.200622380 (accessed on 30th July 2025).

11. Anisimov, S. I., Zhakhovsky, V. V., Inogamov, N. A., Migdal, K. P., Petrov, Yu. V., & Khokhlov, V. A. (2019). High-energy-density physics and laser technologies. Journal of Experimental and Theoretical Physics, 129(4), 757–782. https://doi.org/10.1134/S1063776119100169 (accessed on 30th July 2025).

12. Arnault, J. C., Saada, S., & Ralchenko, V. (2022). Chemical vapor deposition single-crystal diamond: A review. Physica Status Solidi (RRL)–Rapid Research Letters, 16(1), 2100354. https://doi.org/10.1002/pssr.202100354 (accessed on 30th July 2025).

13. Aronsson, B. O., Lausmaa, J., & Kasemo, B. (1997). Glow discharge plasma treatment for surface cleaning and modification of metallic biomaterials. Journal of Biomedical Materials Research, 35(1), 49–73. https://doi.org/10.1002/(SICI)1097-4636(199704)35:1<49::AID-JBM6>3.0.CO;2-M (accessed on 30th July 2025).

14. Arora, P., & Sen Gupta, A. (2023). Atomic frequency standards. In Handbook of Metrology and Applications (pp. 431–453). Springer, Singapore. https://doi.org/10.1007/978-981-99-2074-7_21 (accessed on 30th July 2025).

15. Asryan, L. V., & Suris, R. A. (2004). Theory of threshold characteristics of semiconductor quantum dot lasers. Semiconductors, 38(1), 1–22. https://doi.org/10.1134/1.1641126 (accessed on 30th July 2025).

16. Avazpour, M., Beltrán-Pérez, G., Rodríguez-Morales, L. A., Armas-Rivera, I., Ibarra-Escamilla, B., Muñoz-Aguirre, S., Castillo-Mixcóatl, J., Pottiez, O., & Kuzin, E. A. (2019). The use of polarization-imbalanced NOLM to improve the quality of the spectrum compression. Optics & Laser Technology, 120, 105692. https://doi.org/10.1016/j.optlastec.2019.105692 (accessed on 30th July 2025).

17. Babichev, A., Blokhin, S., Kolodeznyi, E., Karachinsky, L., Novikov, I., Egorov, A., Tian, S. C., & Bimberg, D. (2023). Long-wavelength VCSELs: Status and prospects. Photonics, 10(3), Article 3. https://doi.org/10.3390/photonics10030268 (accessed on 30th July 2025).

18. Balakier, K., Ponnampalam, L., Fice, M. J., Renaud, C. C., & Seeds, A. J. (2018). Integrated semiconductor laser optical phase lock loops. IEEE Journal of Selected Topics in Quantum Electronics, 24(1), 1–12. https://doi.org/10.1109/JSTQE.2017.2711581 (accessed on 30th July 2025).

19. Banerjee, A. (2024). Semiconductor optoelectronic devices. In A. Banerjee (Ed.), Semiconductor Devices: Diodes, Transistors, Solar Cells, Charge Coupled Devices and Solid State Lasers (pp. 245–274). Springer Nature Switzerland. https://doi.org/10.1007/978-3-031-45750-0_14 (accessed on 30th July 2025).

20. Banerjee, P., & Matsakis, D. (2023). Introduction to atomic clocks. In P. Banerjee & D. Matsakis (Eds.), An Introduction to Modern Timekeeping and TIme Transfer (pp. 47–78). Springer Nature Switzerland. https://doi.org/10.1007/978-3-031-30780-5_3 (accessed on 30th July 2025).

21. Bayrakli, I. (2021). Electromagnetically induced transparency in natural and artificial molecules. Optics & Laser Technology, 141, 107168. https://doi.org/10.1016/j.optlastec.2021.107168 (accessed on 30th July 2025).

22. Bedoya-Pinto, A., Chang, K., Samant, M. G., & Parkin, S. (2021). Material preparation/thin film growth. In Handbook of Magnetism and Magnetic Materials (pp. 1–50). Springer, Cham. https://doi.org/10.1007/978-3-030-63101-7_23-1 (accessed on 30th July 2025).

23. Benzarti, Z., Khalfallah, A., Bougrioua, Z., Evaristo, M., & Cavaleiro, A. (2023). Understanding the influence of physical properties on the mechanical characteristics of Mg-doped GaN thin films. Materials Chemistry and Physics, 307, 128182. https://doi.org/10.1016/j.matchemphys.2023.128182 (accessed on 30th July 2025).

24. Bhusal, L., & Mascarenhas, A. (2012). Solid-state lighting. In Handbook of Luminescent Semiconductor Materials. CRC Press.

25. Bimberg, D., Grundmann, M., Heinrichsdorff, F., Ledentsov, N. N., Ustinov, V. M., Zhukov, A. E., Kovsh, A. R., Maximov, M. V., Shernyakov,

Y. M., Volovik, B. V., Tsatsul'nikov, A. F., Kop'ev, P. S., & Alferov, Zh. I. (2000). Quantum dot lasers: Breakthrough in optoelectronics. Thin Solid Films, 367(1), 235–249. https://doi.org/10.1016/S0040-6090(00)00697-0 (accessed on 30th July 2025).

26. Bimberg, D., Grundmann, M., Ledentsov, N. N., Mao, M. H., Ribbat, Ch., Sellin, R., Ustinov, V. M., Zhukov, A. E., Alferov, Zh. I., & Lott, J. A. (2001). Novel infrared quantum dot lasers: Theory and reality. Physica Status Solidi (b), 224(3), 787–796. https://doi.org/10.1002/(SICI)1521-3951(200104)224:3<787::AID-PSSB787>3.0.CO;2-M (accessed on 30th July 2025).

27. Bisquert, J., Garcia-Belmonte, G., & Fabregat-Santiago, F. (1999). Modelling the electric potential distribution in the dark in nanoporous semiconductor electrodes. Journal of Solid-State Electrochemistry, 3(6), 337–347. https://doi.org/10.1007/s100080050164 (accessed on 30th July 2025).

28. Biswas, D. J. (2023a). Different types of lasers. In D. J. Biswas (Ed.), A Beginner's Guide to Lasers and Their Applications, Part 1: Insights into Laser Science (pp. 217–260). Springer International Publishing. https://doi.org/10.1007/978-3-031-24330-1_10 (accessed on 30th July 2025).

29. Biswas, D. J. (2023b). Semiconductor lasers. In D. J. Biswas (Ed.), A Beginner's Guide to Lasers and Their Applications, Part 1: Insights into Laser Science (pp. 287–312). Springer International Publishing. https://doi.org/10.1007/978-3-031-24330-1_12 (accessed on 30th July 2025).

30. Boriskina, S. V., Benson, T. M., Sewell, P., & Nosich, A. I. (2006). Q factor and emission pattern control of the WG modes in notched microdisk resonators. IEEE Journal of Selected Topics in Quantum Electronics, 12(1), 52–58. https://doi.org/10.1109/JSTQE.2005.863002 (accessed on 30th July 2025).

31. Botez, D., & Connolly, J. C. (1981). Single-mode positive-index guided cw constricted double-heterojunction large-optical-cavity AlGaAs lasers with low threshold-current temperature sensitivity. Applied Physics Letters, 38(9), 658–660. https://doi.org/10.1063/1.92495 (accessed on 30th July 2025).

32. Boukortt, N. E. I., Patanè, S., & Abdulraheem, Y. M. (2020). Numerical investigation of CIGS thin-film solar cells. Solar Energy, 204, 440–447. https://doi.org/10.1016/j.solener.2020.05.021 (accessed on 30th July 2025).

33. Brenier, A., & Boulon, G. (2001). Overview of the best Yb^{3+}-doped laser crystals. Journal of Alloys and Compounds, 323, 324, 210–213. https://doi.org/10.1016/S0925-8388(01)01112-4 (accessed on 30th July 2025).

34. Brown, E. R., Globisch, B., Carpintero, G., Rivera, A., Segovia-Vargas, D., & Steiger, A. (2021). Photoconductive THz sources driven at 1550 nm. In Fundamentals ofTerahertz Devices and Applications (pp. 43–136). John Wiley & Sons, Ltd. https://doi.org/10.1002/9781119460749.ch3 (accessed on 30th July 2025).

35. Brown, M. S., & Arnold, C. B. (2010). Fundamentals of laser-material interaction and application to multiscale surface modification. In K. Sugioka, M. Meunier, & A. Piqué (Eds.), Laser Precision Microfabrication (pp. 91–120). Springer. https://doi.org/10.1007/978-3-642-10523-4_4 (accessed on 30th July 2025).

36. Bulashevich, K. A., Mymrin, V. F., Karpov, S. Y., Demidov, D. M., & Ter-Martirosyan, A. L. (2007). Effect of free-carrier absorption on performance of 808 nm AlGaAs-based high-power laser diodes. Semiconductor Science and Technology, 22(5), 502. https://doi.org/10.1088/0268-1242/22/5/008 (accessed on 30th July 2025).

37. Burdov, V. A., & Vasilevskiy, M. I. (2021). Exciton-photon interactions in semiconductor nanocrystals: Radiative transitions, non-radiative processes and environment effects. Applied Sciences, 11(2), Article 2. https://doi.org/10.3390/app11020497 (accessed on 30th July 2025).

38. Cao, H., & Wiersig, J. (2015). Dielectric microcavities: Model systems for wave chaos and non-Hermitian physics. Reviews of Modern Physics, 87(1), 61–111. https://doi.org/10.1103/RevModPhys.87.61 (accessed on 30th July 2025).

39. Cao, X. A., Abernathy, C. R., Singh, R. K., Pearton, S. J., Fu, M., Sarvepalli, V., Sekhar, J. A., Zolper, J. C., Rieger, D. J., Han, J., Drummond, T. J., Shul, R. J., & Wilson, R. G. (1998). Ultrahigh Si^+ implant activation efficiency in GaN using a high-temperature rapid thermal process system. Applied Physics Letters, 73(2), 229–231. https://doi.org/10.1063/1.121764 (accessed on 30th July 2025).

40. Caprioglio, P., Stolterfoht, M., Wolff, C. M., Unold, T., Rech, B., Albrecht, S., & Neher, D. (2019). On the relation between the open-circuit voltage and quasi-Fermi level splitting in efficient perovskite solar cells. Advanced Energy Materials, 9(33), 1901631. https://doi.org/10.1002/aenm.201901631 (accessed on 30th July 2025).

41. Cheng, H., Feng, Y., Fu, Y., Zheng, Y., Shao, Y., & Bai, Y. (2022). Understanding and minimizing non-radiative recombination losses in perovskite light-emitting diodes. Journal of Materials Chemistry C, 10(37), 13590–13610. https://doi.org/10.1039/D2TC01869A (accessed on 30th July 2025).

42. Cheng, K. Y. (2020a). Heterostructure photonic devices. In K. Y. Cheng (Ed.), III–V Compound Semiconductors and Devices: An Introduction to Fundamentals (pp. 419–514). Springer International Publishing. https://doi.org/10.1007/978-3-030-51903-2_10 (accessed on 30th July 2025).

43. Cheng, K. Y. (2020b). Optical properties of compound semiconductor heterostructures. In K. Y. Cheng (Ed.), III–V Compound Semiconductors and Devices: An Introduction to Fundamentals (pp. 289–337). Springer International Publishing. https://doi.org/10.1007/978-3-030-51903-2_8 (accessed on 30th July 2025).

44. Cheng, Y., Yang, Z., Gu, Z., Sun, D., Chen, Y., Guo, H., & Zhang, C. (2025). A review of the development of all-electronic terahertz communication technology based on amplitude shift keying. In C. Chang, F. Qi, L. Zhang, & L. Hou (Eds.), Proceedings of the 2025 China National Conference on Terahertz Biophysics (pp. 539–550). Springer Nature. https://doi.org/10.1007/978-981-96-4886-3_77 (accessed on 30th July 2025).

45. Chiasera, A., Dumeige, Y., Féron, P., Ferrari, M., Jestin, Y., Nunzi Conti, G., Pelli, S., Soria, S., & Righini, G. C. (2010). Spherical whispering-gallery-mode micro resonators. Laser & Photonics Reviews, 4(3), 457–482. https://doi.org/10.1002/lpor.200910016 (accessed on 30th July 2025).

46. Chichibu, S. F., Uedono, A., Kojima, K., Ikeda, H., Fujito, K., Takashima, S., Edo, M., Ueno, K., & Ishibashi, S. (2018). The origins and properties of intrinsic nonradiative recombination centers in wide bandgap GaN and AlGaN. Journal of Applied Physics, 123(16), 161413. https://doi.org/10.1063/1.5012994 (accessed on 30th July 2025).

47. Chow, W. W., & Jahnke, F. (2013). On the physics of semiconductor quantum dots for applications in lasers and quantum optics. Progress in Quantum Electronics, 37(3), 109–184. https://doi.org/10.1016/j.pquantelec.2013.04.001 (accessed on 30th July 2025).

48. Chui, H., Gardner, N. F., Grillot, P. N., Huang, J. W., Krames, M. R., & Maranowski, S. A. (1999). Chapter 2 high-efficiency AlGaInP light-emitting diodes. In G. Mueller (Ed.), Semiconductors and Semimetals (Vol. 64, pp. 49–128). Elsevier. https://doi.org/10.1016/S0080-8784(08)62488-6 (accessed on 30th July 2025).

49. Cohen-Tannoudji, C. N. (1998). Nobel lecture: Manipulating atoms with photons. Reviews of Modern Physics, 70(3), 707–719. https://doi.org/10.1103/RevModPhys.70.707 (accessed on 30th July 2025).

50. Cohen-Tannoudji, C. N., & Phillips, W. D. (1990). New mechanisms for laser cooling. Physics Today, 43(10), 33–40. https://doi.org/10.1063/1.881239 (accessed on 30th July 2025).

51. Connelly, M. J. (Ed.). (2002). Functional applications. In Semiconductor Optical Amplifiers (pp. 127–165). Springer US. https://doi.org/10.1007/0-306-48156-1_7 (accessed on 30th July 2025).

52. Costales, A., Blanco, M. A., Martín Pendás, Á., Kandalam, A. K., & Pandey, R. (2002). Chemical bonding in group III nitrides. Journal of the American Chemical Society, 124(15), 4116–4123. https://doi.org/10.1021/ja017380o (accessed on 30th July 2025).

53. Cowan, S. R., Wang, J., Yi, J., Lee, Y. J., Olson, D. C., & Hsu, J. W. P. (2013). Intensity and wavelength dependence of bimolecular recombination in P3HT:PCBM solar cells: A white-light biased external quantum efficiency study. Journal of Applied Physics, 113(15), 154504. https://doi.org/10.1063/1.4801920 (accessed on 30th July 2025).

54. Dalir, H., Matsutani, A., Ahmed, M., Bakry, A., & Koyama, F. (2014). High frequency modulation of transverse-coupled-cavity VCSELs for radio over fiber applications. IEEE Photonics Technology Letters, 26(3), 281–284. https://doi.org/10.1109/LPT.2013.2292577 (accessed on 30th July 2025).

55. Datta, R., Kappers, M. J., Vickers, M. E., Barnard, J. S., & Humphreys, C. J. (2004). Growth and characterization of GaN with reduced dislocation density. Superlattices and Microstructures, 36(4), 393–401. https://doi.org/10.1016/j.spmi.2004.09.003 (accessed on 30th July 2025).

56. de Arruda Mello, D. A., & Barbosa, F. A. (2021). The optical transmitter. In D. A. de Arruda Mello & F. A. Barbosa (Eds.), Digital Coherent Optical Systems: Architecture and Algorithms (pp. 19–46). Springer International Publishing. https://doi.org/10.1007/978-3-030-66541-8_2 (accessed on 30th July 2025).

57. de Melo, O. (2023). Introduction in II–VI semiconductors. In G. Korotcenkov (Ed.), Handbook of II–VI Semiconductor-Based Sensors and Radiation Detectors: Volume 1, Materials and technology (pp. 3–19). Springer International Publishing. https://doi.org/10.1007/978-3-031-19531-0_1 (accessed on 30th July 2025).

58. DenBaars, S. P., Feezell, D., Kelchner, K., Pimputkar, S., Pan, C. C., Yen, C. C., Tanaka, S., Zhao, Y., Pfaff, N., Farrell, R., Iza, M., Keller, S., Mishra, U., Speck, J. S., & Nakamura, S. (2013). Development of gallium-nitride-based light-emitting diodes (LEDs) and laser diodes for energy-efficient lighting and displays. Acta Materialia, 61(3), 945–951. https://doi.org/10.1016/j.actamat.2012.10.042 (accessed on 30th July 2025).

59. Diddams, S. A., Bergquist, J. C., Jefferts, S. R., & Oates, C. W. (2004). Standards of time and frequency at the outset of the 21st century.

Science, 306(5700), 1318–1324. https://doi.org/10.1126/science.1102330 (accessed on 30th July 2025).

60. Dong, H., Sun, J., Ma, S., Liang, J., Lu, T., Jia, Z., Liu, X., & Xu, B. (2016). Effect of potential barrier height on the carrier transport in InGaAs/GaAsP multi-quantum wells and photoelectric properties of laser diode. Physical Chemistry Chemical Physics, 18(9), 6901–6912. https://doi.org/10.1039/C5CP07805A (accessed on 30th July 2025).
61. Dowell, D. H., & Schmerge, J. F. (2009). Quantum efficiency and thermal emittance of metal photocathodes. Physical Review Special Topics–Accelerators and Beams, 12(7), 074201. https://doi.org/10.1103/PhysRevSTAB.12.074201 (accessed on 30th July 2025).
62. Dutta, N. K. (2002). Chapter 2—Long-wavelength laser source. In A. K. Dutta, N. K. Dutta, & M. Fujiwara (Eds.), WDM Technologies (pp. 13–58). Academic Press. https://doi.org/10.1016/B978-012225261-7/50004-X (accessed on 30th July 2025).
63. Dutton, R. W., & Yu, Z. (1993). PN junctions. In R. W. Dutton & Z. Yu (Eds.), Technology CAD–Computer Simulation of IC Processes and Devices (pp. 131–195). Springer US. https://doi.org/10.1007/978-1-4615-3208-8_4 (accessed on 30th July 2025).
64. Edmond, J., Abare, A., Bergman, M., Bharathan, J., Lee Bunker, K., Emerson, D., Haberern, K., Ibbetson, J., Leung, M., Russel, P., & Slater, D. (2004). High efficiency GaN-based LEDs and lasers on SiC. Journal of Crystal Growth, 272(1), 242–250. https://doi.org/10.1016/j.jcrysgro.2004.08.056 (accessed on 30th July 2025).
65. Eichler, H. J., Eichler, J., & Lux, O. (2018). Semiconductor lasers. In H. J. Eichler, J. Eichler, & O. Lux (Eds.), Lasers: Basics, Advances and Applications (pp. 165–203). Springer International Publishing. https://doi.org/10.1007/978-3-319-99895-4_10 (accessed on 30th July 2025).
66. Eliseev, P. G. (1980). Stripe-geometry lasers and their properties. Czechoslovak Journal of Physics B, 30(3), 300–317. https://doi.org/10.1007/BF01596256 (accessed on 30th July 2025).
67. El-Kareh, B., & Hutter, L. N. (2024). Silicon Components and Processes Self Study: Unit Processes and Process Integration. Springer Nature Switzerland. https://doi.org/10.1007/978-3-031-59219-5 (accessed on 30th July 2025).
68. Evstigneev, M. (2022). Metal–semiconductor contact. In M. Evstigneev (Ed.), Introduction to Semiconductor Physics and Devices (pp. 215–232). Springer International Publishing. https://doi.org/10.1007/978-3-031-08458-4_9 (accessed on 30th July 2025).

69. Fabis, P. M. (1999). Material property, compatibility, and reliability issues in diamond-enhanced, GaAs-based plastic packages. Microelectronics Reliability, 39(8), 1275–1291. https://doi.org/10.1016/S0026-2714(99)00040-2 (accessed on 30th July 2025).

70. Fabis, P. M. (2002). The processing technology and electronic packaging of CVD diamond: A case study for GaAs/CVD diamond plastic packages. Microelectronics Reliability, 42(2), 233–252. https://doi.org/10.1016/S0026-2714(01)00242-6 (accessed on 30th July 2025).

71. Faillon, G., Kornfeld, G., Bosch, E., & Thumm, M. K. (2008). Microwave tubes. In J. A. Eichmeier & M. K. Thumm (Eds.), Vacuum Electronics: Components and Devices (pp. 1–84). Springer. https://doi.org/10.1007/978-3-540-71929-8_1 (accessed on 30th July 2025).

72. Fan, X., Shi, J., Chen, Y., Miao, G., Jiang, H., & Song, H. (2024). A comprehensive review of group-III nitride light-emitting diodes: From millimeter to micro-nanometer scales. Micromachines, 15(10), Article 10. https://doi.org/10.3390/mi15101188 (accessed on 30th July 2025).

73. Fang, Z. C., Wu, Z. L., Huang, C. G., & Wu, C. W. (2022). Micro-scale thermodynamic model of microstructure and stress evolution in parts via selective laser melting. Journal of Materials Science, 57(25), 11918–11933. https://doi.org/10.1007/s10853-022-07046-6 (accessed on 30th July 2025).

74. Fang, Z., Cai, H., Chen, G., & Qu, R. (2017a). Applications of single-frequency semiconductor lasers. In Z. Fang, H. Cai, G. Chen, & R. Qu (Eds.), Single Frequency Semiconductor Lasers (pp. 267–302). Springer. https://doi.org/10.1007/978-981-10-5257-6_9 (accessed on 30th July 2025).

75. Fang, Z., Cai, H., Chen, G., & Qu, R. (2017b). Monolithically integrated semiconductor lasers. In Z. Fang, H. Cai, G. Chen, & R. Qu (Eds.), Single Frequency Semiconductor Lasers (pp. 81–115). Springer. https://doi.org/10.1007/978-981-10-5257-6_4 (accessed on 30th July 2025).

76. Fathpour, S., Mi, Z., & Bhattacharya, P. (2005). High-speed quantum dot lasers. Journal of Physics D: Applied Physics, 38(13), 2103. https://doi.org/10.1088/0022-3727/38/13/005 (accessed on 30th July 2025).

77. Fiorini, C., Gianoncelli, A., Longoni, A., & Zaraga, F. (2002). Determination of the thickness of coatings by means of a new XRF spectrometer. X-Ray Spectrometry, 31(1), 92–99. https://doi.org/10.1002/xrs.550 (accessed on 30th July 2025).

78. Fischer, A. J., Allerman, A. A., Crawford, M. H., Bogart, K. H. A., Lee, S. R., Kaplar, R. J., Chow, W. W., Kurtz, S. R., Fullmer, K. W., & Figiel,

J. J. (2004). Room-temperature direct current operation of 290 nm light-emitting diodes with milliwatt power levels. Applied Physics Letters, 84(17), 3394–3396. https://doi.org/10.1063/1.1728307 (accessed on 30th July 2025).

79. Fleischmann, T., Moran, M., Hopkinson, M., Meidia, H., Rees, G. J., Cullis, A. G., Sánchez-Rojas, J. L., & Izpura, I. (2001). Strained layer (111)B GaAs/InGaAs single quantum well lasers and the dependence of their characteristics upon indium composition. Journal of Applied Physics, 89(9), 4689–4696. https://doi.org/10.1063/1.1359155 (accessed on 30th July 2025).

80. Fox, M., & Ispasoiu, R. (2017). Quantum wells, superlattices, and band-gap engineering. In S. Kasap & P. Capper (Eds.), Springer Handbook of Electronic and Photonic Materials (pp. 1–1). Springer International Publishing. https://doi.org/10.1007/978-3-319-48933-9_40 (accessed on 30th July 2025).

81. Gamaly, E. G., & Rode, A. V. (2013). Physics of ultra-short laser interaction with matter: From phonon excitation to ultimate transformations. Progress in Quantum Electronics, 37(5), 215–323. https://doi.org/10.1016/j.pquantelec.2013.05.001 (accessed on 30th July 2025).

82. Gholipour, S., Rahighi, R., Panahi, M., Mirsepah, A., Bakhshayesh, A., Shahbaz Tehrani, F., Pourjafarabad, M., Ahmed, F., Ansari, M. Z., & Abdi, Y. (2024). Photodiodes and phototransistors based on nanomaterials-silicon heterostructures: A review. Optical and Quantum Electronics, 57(1), 23. https://doi.org/10.1007/s11082-024-07119-7 (accessed on 30th July 2025).

83. Giesecke, G. (1966). Chapter 4 lattice constants. In R. K. Willardson & A. C. Beer (Eds.), Semiconductors and Semimetals (Vol. 2, pp. 63–73). Elsevier. https://doi.org/10.1016/S0080-8784(08)60162-3 (accessed on 30th July 2025).

84. Goodrich, A., Hacke, P., Wang, Q., Sopori, B., Margolis, R., James, T. L., & Woodhouse, M. (2013). A wafer-based monocrystalline silicon photovoltaics road map: Utilizing known technology improvement opportunities for further reductions in manufacturing costs. Solar Energy Materials and Solar Cells, 114, 110–135. https://doi.org/10.1016/j.solmat.2013.01.030 (accessed on 30th July 2025).

85. Gravrand, O., Rothman, J., Delacourt, B., Boulard, F., Lobre, C., Ballet, Ph., Santailler, J. L., Cervera, C., Brellier, D., Péré-Laperne, N., Destefanis, V., & Kerlain, A. (2018). Shockley–Read–Hall lifetime study and implication in HgCdTe photodiodes for IR detection. Journal of Electronic Materials,

47(10), 5680–5690. https://doi.org/10.1007/s11664-018-6557-2 (accessed on 30th July 2025).

86. Gribkovskii, V. P. (1995). Injection lasers. Progress in Quantum Electronics, 19(1), 41–88. https://doi.org/10.1016/0079-6727(94)E0005-I (accessed on 30th July 2025).

87. Gryga, M., Ciprian, D., & Hlubina, P. (2022). Distributed Bragg reflectors employed in sensors and filters based on cavity-mode spectral-domain resonances. Sensors, 22(10), Article 10. https://doi.org/10.3390/s22103627 (accessed on 30th July 2025).

88. Gunnarsson, W. B., Roh, K., Zhao, L., Murphy, J. P., Grede, A. J., Giebink, N. C., & Rand, B. P. (2023). Toward nonepitaxial laser diodes. Chemical Reviews, 123(12), 7548–7584. https://doi.org/10.1021/acs.chemrev.2c00721 (accessed on 30th July 2025).

89. Guo, N., Gong, F., Liu, J., Jia, Y., Zhao, S., Liao, L., Su, M., Fan, Z., Chen, X., Lu, W., Xiao, L., & Hu, W. (2017). Hybrid WSe_2–In_2O_3 phototransistor with ultrahigh detectivity by efficient suppression of dark currents. ACS Applied Materials & Interfaces, 9(39), 34489–34496. https://doi.org/10.1021/acsami.7b10698 (accessed on 30th July 2025).

90. Gupta, K. M., & Gupta, N. (2016). Optoelectronic devices. In K. M. Gupta & N. Gupta (Eds.), Advanced Semiconducting Materials and Devices (pp. 311–350). Springer International Publishing. https://doi.org/10.1007/978-3-319-19758-6_9 (accessed on 30th July 2025).

91. Haken, H. (1966). Theory of intensity and phase fluctuations of a homogeneously broadened laser. Zeitschrift für Physik, 190(3), 327–356. https://doi.org/10.1007/BF01333598 (accessed on 30th July 2025).

92. Hänsch, T. W., & Picqué, N. (2012). Frequency combs. In F. Träger (Ed.), Springer Handbook of Lasers and Optics (pp. 1285–1304). Springer. https://doi.org/10.1007/978-3-642-19409-2_17 (accessed on 30th July 2025).

93. Hardtdegen, H., & Mikulics, M. (2019). Column III. In Metalorganic Vapor Phase Epitaxy (MOVPE) (pp. 71–108). John Wiley & Sons, Ltd. https://doi.org/10.1002/9781119313021.ch3 (accessed on 30th July 2025).

94. Hardy, M. T., Feezell, D. F., DenBaars, S. P., & Nakamura, S. (2011). Group III-nitride lasers: A materials perspective. Materials Today, 14(9), 408–415. https://doi.org/10.1016/S1369-7021(11)70185-7 (accessed on 30th July 2025).

95. He, L., Özdemir, Ş. K., & Yang, L. (2013). Whispering gallery microcavity lasers. Laser & Photonics Reviews, 7(1), 60–82. https://doi.org/10.1002/lpor.201100032 (accessed on 30th July 2025).

96. Hertkorn, J., Brückner, P., Thapa, S. B., Wunderer, T., Scholz, F., Feneberg, M., Thonke, K., Sauer, R., Beer, M., & Zweck, J. (2007). Optimization of nucleation and buffer layer growth for improved GaN quality. Journal of Crystal Growth, 308(1), 30–36. https://doi.org/10.1016/j.jcrysgro.2007.07.056 (accessed on 30th July 2025).

97. Hipp, D., Mahrle, A., & Beyer, E. (2019). Beyond Fresnel: Absorption of fiber laser radiation on rough stainless-steel surfaces. Journal of Physics D: Applied Physics, 52(35), 355302. https://doi.org/10.1088/1361-6463/ab2415 (accessed on 30th July 2025).

98. Hjort, K. (2004). Transfer of InP epilayers by wafer bonding. Journal of Crystal Growth, 268(3), 346–358. https://doi.org/10.1016/j.jcrysgro.2004.04.053 (accessed on 30th July 2025).

99. Hoffmann, A. (1996). Optical properties of GaN. In R. Helbig (Ed.), Festkörperprobleme 36 (pp. 33–56). Springer. https://doi.org/10.1007/BFb0107667 (accessed on 30th July 2025).

100. Holland, P. R. (1993). The de Broglie-Bohm theory of motion and quantum field theory. Physics Reports, 224(3), 95–150. https://doi.org/10.1016/0370-1573(93)90095-U (accessed on 30th July 2025).

101. Hu, F. Q. (2004). Absorbing boundary conditions. International Journal of Computational Fluid Dynamics, 18(6), 513–522. https://doi.org/10.1080/10618560410001673524 (accessed on 30th July 2025).

102. Hubbard, S. (2016). Recombination. In Photovoltaic Solar Energy (pp. 39–46). John Wiley & Sons, Ltd. https://doi.org/10.1002/9781118927496.ch5 (accessed on 30th July 2025).

103. Huebener, R. P. (2016). Less can be more: Semiconductors. In R. P. Huebener (Ed.), Conductors, Semiconductors, Superconductors: An Introduction to Solid State Physics (pp. 73–96). Springer International Publishing. https://doi.org/10.1007/978-3-319-24010-7_6 (accessed on 30th July 2025).

104. Ji, H., & Abshire, P. A. (2004). Fundamentals of silicon-based phototransduction. In O. Yadid-Pecht & R. Etienne-Cummings (Eds.), CMOS Imagers: From Phototransduction to Image Processing (pp. 1–51). Springer US. https://doi.org/10.1007/1-4020-7962-1_1 (accessed on 30th July 2025).

105. Jiang, G., Diao, L., & Kuang, K. (2013). Understanding lasers, laser diodes, laser diode packaging and their relationship to tungsten copper. In G. Jiang, L. Diao, & K. Kuang (Eds.), Advanced Thermal Management Materials (pp. 123–140). Springer. https://doi.org/10.1007/978-1-4614-1963-1_9 (accessed on 30th July 2025).

106. Jiang, W., Nolan, M., Ehrentraut, D., & D'Evelyn, M. P. (2017). Electrical and optical properties of gallium vacancy complexes in ammonothermal GaN. Applied Physics Express, 10(7), 075506. https://doi.org/10.7567/APEX.10.075506 (accessed on 30th July 2025).

107. Jirauschek, C., & Kubis, T. (2014). Modeling techniques for quantum cascade lasers. Applied Physics Reviews, 1(1), 011307. https://doi.org/10.1063/1.4863665 (accessed on 30th July 2025).

108. Kaplan, I. G. (1992). Exclusion principle and indistinguishability of identical particles in quantum mechanics. Journal of Molecular Structure, 272, 187–196. https://doi.org/10.1016/0022-2860(92)80032-D (accessed on 30th July 2025).

109. Kidalov, S. V., & Shakhov, F. M. (2009). Thermal conductivity of diamond composites. Materials, 2(4), Article 4. https://doi.org/10.3390/ma2042467 (accessed on 30th July 2025).

110. Kim, S., Park, J., Kim, S., Lee, Y., Ahn, S., Cho, J., Ju, M., Lakshminarayan, N., Dao, V. A., & Yi, J. (2014). Free-carrier absorption and Burstein–Moss shift effect on quantum efficiency in heterojunction silicon solar cells. Vacuum, 108, 39–44. https://doi.org/10.1016/j.vacuum.2014.05.015 (accessed on 30th July 2025).

111. Kippenberg, T. J., Spillane, S. M., & Vahala, K. J. (2004). Demonstration of ultra-high-Q small mode volume toroid microcavities on a chip. Applied Physics Letters, 85(25), 6113–6115. https://doi.org/10.1063/1.1833556 (accessed on 30th July 2025).

112. Klotzkin, D. J. (2014a). Semiconductor laser operation. In D. J. Klotzkin (Ed.), Introduction to Semiconductor Lasers for Optical Communications: An Applied Approach (pp. 81–107). Springer. https://doi.org/10.1007/978-1-4614-9341-9_5 (accessed on 30th July 2025).

113. Klotzkin, D. J. (2014b). Semiconductors as laser material 1: Fundamentals. In D. J. Klotzkin (Ed.), Introduction to Semiconductor Lasers for Optical Communications: An Applied Approach (pp. 31–52). Springer. https://doi.org/10.1007/978-1-4614-9341-9_3 (accessed on 30th July 2025).

114. Koechner, W. (1996). Optical pump systems. In W. Koechner (Ed.), Solid-State Laser Engineering (pp. 281–392). Springer. https://doi.org/10.1007/978-3-662-14105-2_6 (accessed on 30th July 2025).

115. Kojima, K., & Kyuma, K. (1990). Multi-quantum well distributed feedback and distributed Bragg reflector lasers. Semiconductor Science and Technology, 5(6), 481. https://doi.org/10.1088/0268-1242/5/6/003 (accessed on 30th July 2025).

116. Komljenovic, T., Liang, L., Chao, R. L., Hulme, J., Srinivasan, S., Davenport, M., & Bowers, J. E. (2017). Widely-tunable ring-resonator semiconductor lasers. Applied Sciences, 7(7), Article 7. https://doi.org/10.3390/app7070732 (accessed on 30th July 2025).

117. Koyama, F. (2014). Advances and new functions of VCSEL photonics. Optical Review, 21(6), 893–904. https://doi.org/10.1007/s10043-014-0142-6 (accessed on 30th July 2025).

118. Kuhl, J. (1993). Contact-free characterization of electronic and optoelectronic devices with ultrashort laser pulses. In W. E. Bron (Ed.), Ultrashort Processes in Condensed Matter (pp. 143–195). Springer US. https://doi.org/10.1007/978-1-4615-2954-5_4 (accessed on 30th July 2025).

119. Kumar, A., & Krishnan, V. (2021). Vacancy engineering in semiconductor photocatalysts: Implications in hydrogen evolution and nitrogen fixation applications. Advanced Functional Materials, 31(28), 2009807. https://doi.org/10.1002/adfm.202009807 (accessed on 30th July 2025).

120. Lagraa, I., Soudini, B., Abid, H., & Taleb, S. (2022). Study and optimization of structure InAs/InGaAs quantum dot in-a-well long-wave infrared photodetector. Optik, 251, 168494. https://doi.org/10.1016/j.ijleo.2021.168494 (accessed on 30th July 2025).

121. Le, H. Q., Turner, G. W., Ochoa, J. R., Manfra, M. J., Cook, C. C., & Zhang, Y. (1997). External cavity mid-infrared semiconductor lasers. In-Plane Semiconductor Lasers: From Ultraviolet to Midinfrared, 3001, 298–308. https://doi.org/10.1117/12.273799 (accessed on 30th July 2025).

122. Lee, H., Chen, T., Li, J., Yang, K. Y., Jeon, S., Painter, O., & Vahala, K. J. (2012). Chemically etched ultrahigh-Q wedge-resonator on a silicon chip. Nature Photonics, 6(6), 369–373. https://doi.org/10.1038/nphoton.2012.109 (accessed on 30th July 2025).

123. Li, H., Li, D., Lou, Q., Liu, C., Lan, T., & Yu, X. (2024). Advancements in optical resonator stability: Principles, technologies, and applications. Sensors, 24(19), Article 19. https://doi.org/10.3390/s24196473 (accessed on 30th July 2025).

124. Li, S. S. (2006). Light-emitting devices. In S. S. Li (Ed.), Semiconductor Physical Electronics (pp. 458–512). Springer. https://doi.org/10.1007/0-387-37766-2_13 (accessed on 30th July 2025).

125. Li, X., Zhang, Y., Wang, J., Xiong, L., Zhang, P., Nie, Z., Wang, Z., Liu, H., & Liu, X. (2012). Influence of package structure on the performance of the single emitter diode laser. IEEE Transactions on Components, Packaging and Manufacturing Technology, 2(10), 1592–1599. https://doi.

org/10.1109/TCPMT.2012.2207456 (accessed on 30th July 2025).

126. Liang, J., Dariavach, N., & Shangguan, D. (2007). Metallurgy, processing and reliability of lead-free solder joint interconnections. In E. Suhir, Y. C. Lee, & C. P. Wong (Eds.), Micro- and Opto-Electronic Materials and Structures: Physics, Mechanics, Design, Reliability, Packaging (pp. A351–A409). Springer US. https://doi.org/10.1007/0-387-32989-7_10 (accessed on 30th July 2025).

127. Liao, Y. S., Kuo, H. C., Chen, Y. J., & Lin, G. R. (2009). Side-mode transmission diagnosis of a multichannel selectable injection-locked Fabry-Perot laser diode with anti-reflection coated front facet. Optics Express, 17(6), 4859–4867. https://doi.org/10.1364/OE.17.004859 (accessed on 30th July 2025).

128. Lin, G., Coillet, A., & Chembo, Y. K. (2017). Nonlinear photonics with high-Q whispering-gallery-mode resonators. Advances in Optics and Photonics, 9(4), 828–890. https://doi.org/10.1364/AOP.9.000828 (accessed on 30th July 2025).

129. Liu, K., Sakurai, M., & Aono, M. (2010). ZnO-based ultraviolet photodetectors. Sensors, 10(9), Article 9. https://doi.org/10.3390/s100908604 (accessed on 30th July 2025).

130. Liu, X. W., Hopgood, A. A., Usher, B. F., Wang, H., & Braithwaite, N. S. J. (1999). Formation of misfit dislocations during growth of In⊠Ga$_1$⊠⊠As/GaAs strained-layer heterostructures. Semiconductor Science and Technology, 14(12), 1154. https://doi.org/10.1088/0268-1242/14/12/325 (accessed on 30th July 2025).

131. Liu, X., Zhao, W., Xiong, L., & Liu, H. (2015a). Introduction of high-power semiconductor lasers. In X. Liu, W. Zhao, L. Xiong, & H. Liu (Eds.), Packaging of High-Power Semiconductor Lasers (pp. 1–28). Springer. https://doi.org/10.1007/978-1-4614-9263-4_1 (accessed on 30th July 2025).

132. Liu, X., Zhao, W., Xiong, L., & Liu, H. (2015b). Overview of high-power semiconductor laser packages. In X. Liu, W. Zhao, L. Xiong, & H. Liu (Eds.), Packaging of High-Power Semiconductor Lasers (pp. 29–52). Springer. https://doi.org/10.1007/978-1-4614-9263-4_2 (accessed on 30th July 2025).

133. Liu, X., Zhao, W., Xiong, L., & Liu, H. (2015c). Packaging process of high-power semiconductor lasers. In X. Liu, W. Zhao, L. Xiong, & H. Liu (Eds.), Packaging of High-Power Semiconductor Lasers (pp. 185–226). Springer. https://doi.org/10.1007/978-1-4614-9263-4_7 (accessed on 30th July 2025).

134. Liu, X., Zhao, W., Xiong, L., & Liu, H. (2015d). Thermal design and management in high power semiconductor laser packaging. In X. Liu, W. Zhao, L. Xiong, & H. Liu (Eds.), Packaging of High-Power Semiconductor Lasers (pp. 53–88). Springer. https://doi.org/10.1007/978-1-4614-9263-4_3 (accessed on 30th July 2025).

135. Lu, Y., Reyes, P. I., Zhong, J., & Chen, H. (2012). Multifunctional ZnO nanostructure-based devices. In S. Pearton (Ed.), GaN and ZnO-Based Materials and Devices (pp. 361–411). Springer. https://doi.org/10.1007/978-3-642-23521-4_13 (accessed on 30th July 2025).

136. Lutz, Y., & Poyet, J. M. (2014). Laser diode stack beam shaping for efficient and compact long-range laser illuminator design. Optics & Laser Technology, 57, 90–95. https://doi.org/10.1016/j.optlastec.2013.09.015 (accessed on 30th July 2025).

137. Ma, D., & Chen, Y. (2017). Physics basis of organic semiconductor heterojunctions. In D. Ma & Y. Chen (Eds.), Organic Semiconductor Heterojunctions and Its Application in Organic Light-Emitting Diodes (pp. 1–36). Springer. https://doi.org/10.1007/978-3-662-53695-7_1 (accessed on 30th July 2025).

138. Ma, X., & Zhong, L. (2008). Advances in high-power semiconductor diode lasers. Semiconductor Lasers and Applications III, 6824, 13–28. https://doi.org/10.1117/12.785610 (accessed on 30th July 2025).

139. Macêdo, A. M. S., González, I. R. R., Raposo, E. P., Menezes, L. de S., & Gomes, A. S. L. (2019). Turbulent intermittency in a random fiber laser. Atoms, 7(2), Article 2. https://doi.org/10.3390/atoms7020043 (accessed on 30th July 2025).

140. Mahamood, R. M. (2018). Laser basics and laser material interactions. In R. M. Mahamood (Ed.), Laser Metal Deposition Process of Metals, Alloys, And Composite Materials (pp. 11–35). Springer International Publishing. https://doi.org/10.1007/978-3-319-64985-6_2 (accessed on 30th July 2025).

141. Malin, T. V., Mansurov, V. G., Galitsyn, Yu. G., Milakhin, D. S., Protasov, D. Yu., Ber, B. Ya., Kazantsev, D. Yu., Ratnikov, V. V., Shcheglov, M. P., Smirnov, A. N., Davydov, V. Yu., & Zhuravlev, K. S. (2021). Mg_3N_2 nano crystallites formation during the GaN:Mg layers growth by the NH_3-MBE technique. Journal of Crystal Growth, 554, 125963. https://doi.org/10.1016/j.jcrysgro.2020.125963 (accessed on 30th July 2025).

142. Maximov, M. V., Nadtochiy, A. M., Mintairov, S. A., Kalyuzhnyy, N. A., Kryzhanovskaya, N. V., Moiseev, E. I., Gordeev, N. Y., Shernyakov, Y. M., Payusov, A. S., Zubov, F. I., Nevedomskiy, V. N., Rouvimov, S. S., &

Zhukov, A. E. (2020). Light emitting devices based on quantum well-dots. Applied Sciences, 10(3), Article 3. https://doi.org/10.3390/app10031038 (accessed on 30th July 2025).

143. Md Taib, M. I., Waheeda, S. N., Jasman, F., Yusop, M. Z. M., & Zainal, N. (2022). GaN nucleation on patterned sapphire substrate with different shapes for improved GaN overgrowth. Vacuum, 197, 110848. https://doi.org/10.1016/j.vacuum.2021.110848 (accessed on 30th July 2025).

144. Michalzik, R. (2013). VCSEL fundamentals. In R. Michalzik (Ed.), VCSELs: Fundamentals, Technology and Applications of Vertical-Cavity Surface-Emitting Lasers (pp. 19–75). Springer. https://doi.org/10.1007/978-3-642-24986-0_2 (accessed on 30th July 2025).

145. Michalzik, R., & Ebeling, K. J. (2003). Operating principles of VCSELs. In H. E. Li & K. Iga (Eds.), Vertical-Cavity Surface-Emitting Laser Devices (pp. 53–98). Springer. https://doi.org/10.1007/978-3-662-05263-1_3 (accessed on 30th July 2025).

146. Milnes, A. G. (1980). Semiconductor junctions and diodes. In A. G. Milnes (Ed.), Semiconductor Devices and Integrated Electronics (pp. 1–84). Springer Netherlands. https://doi.org/10.1007/978-94-011-7021-5_1 (accessed on 30th July 2025).

147. Mishra, U. K., & Singh, J. (2008a). Electronic levels in semiconductors. In U. K. Mishra & J. Singh (Eds.), Semiconductor Device Physics and Design (pp. 28–91). Springer Netherlands. https://doi.org/10.1007/978-1-4020-6481-4_2 (accessed on 30th July 2025).

148. Mishra, U. K., & Singh, J. (2008b). Junctions in semiconductors: P-N diodes. In U. K. Mishra & J. Singh (Eds.), Semiconductor Device Physics and Design (pp. 146–215). Springer Netherlands. https://doi.org/10.1007/978-1-4020-6481-4_4 (accessed on 30th July 2025).

149. Mishra, U. K., & Singh, J. (2008c). Structural properties of semiconductors. In U. K. Mishra & J. Singh (Eds.), Semiconductor Device Physics and Design (pp. 1–27). Springer Netherlands. https://doi.org/10.1007/978-1-4020-6481-4_1 (accessed on 30th July 2025).

150. Miyajima, Y., Nakamura, Y., Konishi, Y., Ishikawa, K., Wang, W., & Takata, N. (2023). Effect of low-temperature annealing on electrical resistivity and mechanical properties of laser-powder bed fused AlSi10Mg alloy. Materials Science and Engineering: A, 871, 144876. https://doi.org/10.1016/j.msea.2023.144876 (accessed on 30th July 2025).

151. Mobini, E., Espinosa, D. H. G., Vyas, K., & Dolgaleva, K. (2022). AlGaAs nonlinear integrated photonics. Micromachines, 13(7), Article 7. https://doi.org/10.3390/mi13070991 (accessed on 30th July 2025).

152. Monemar, B. (2008). Perspective on the development of III-nitrides for optical emitters. Materials Science Forum, 590, 17–26. https://doi.org/10.4028/www.scientific.net/MSF.590.17 (accessed on 30th July 2025).

153. Monier-Vinard, E., Rogie, B., Bissuel, V., Laraqi, N., Daniel, O., & Kotelon, M. C. (2017). State of the art of thermal characterization of electronic components using computational fluid dynamic tools. International Journal of Numerical Methods for Heat & Fluid Flow, 27(11), 2433–2450. https://doi.org/10.1108/HFF-10-2016-0380 (accessed on 30th July 2025).

154. Morkoç, H. (1999). Semiconductor lasers. In H. Morkoç (Ed.), Nitride Semiconductors and Devices (pp. 379–459). Springer. https://doi.org/10.1007/978-3-642-58562-3_12 (accessed on 30th July 2025).

155. Morkoç, H., Strite, S., Gao, G. B., Lin, M. E., Sverdlov, B., & Burns, M. (1994). Large-band-gap SiC, III-V nitride, and II-VI ZnSe-based semiconductor device technologies. Journal of Applied Physics, 76(3), 1363–1398. https://doi.org/10.1063/1.358463 (accessed on 30th July 2025).

156. Murray, A. J., Hussey, M. J., & Needham, M. (2006). Design and characterization of an atomic beam source with narrow angular divergence for alkali-earth targets. Measurement Science and Technology, 17(11), 3094. https://doi.org/10.1088/0957-0233/17/11/030 (accessed on 30th July 2025).

157. Nakamura, S., & Chichibu, S. F. (2000). Introduction to Nitride Semiconductor Blue Lasers and Light Emitting Diodes. CRC Press. https://doi.org/10.1201/9781482268065 (accessed on 30th July 2025).

158. Nakarnura, S. (1997). Chapter 8 Group III-V nitride-based ultraviolet blue-green-yellow light-emitting diodes and laser diodes. In G. B. Stringfellow & M. George Craford (Eds.), Semiconductors and Semimetals (Vol. 48, pp. 391–443). Elsevier. https://doi.org/10.1016/S0080-8784(08)62409-6 (accessed on 30th July 2025).

159. Nakwaski, W., & Sarzał, R. P. (2016). Comprehensive and fully self-consistent modeling of modern semiconductor lasers. Journal of Semiconductors, 37(2), 024001. https://doi.org/10.1088/1674-4926/37/2/024001 (accessed on 30th July 2025).

160. Nawaz, M., & Ahmad, A. (2012). A TCAD-based modeling of GaN/InGaN/Si solar cells. Semiconductor Science and Technology, 27(3), 035019. https://doi.org/10.1088/0268-1242/27/3/035019 (accessed on 30th July 2025).

161. Norman, J. C., Jung, D., Liu, A. Y., Selvidge, J., Mukherjee, K., Bowers, J. E., & Herrick, R. W. (2021). Chapter 6—Reliability of lasers on silicon substrates for silicon photonics. In R. W. Herrick & O. Ueda (Eds.), *Reliability of Semiconductor Lasers and Optoelectronic Devices* (pp. 239–271). Woodhead Publishing. https://doi.org/10.1016/B978-0-12-819254-2.00002-3 (accessed on 30th July 2025).

162. Norman, J. C., Mirin, R. P., & Bowers, J. E. (2021). Quantum dot lasers—History and future prospects. *Journal of Vacuum Science & Technology A, 39*(2), 020802. https://doi.org/10.1116/6.0000768 (accessed on 30th July 2025).

163. Numai, T. (2015). Fundamentals of semiconductor lasers. In T. Numai (Ed.), *Fundamentals of Semiconductor Lasers* (pp. 89–186). Springer Japan. https://doi.org/10.1007/978-4-431-55148-5_5 (accessed on 30th July 2025).

164. Ohtsubo, J. (2017). Semiconductor lasers and theory. In J. Ohtsubo (Ed.), *Semiconductor Lasers: Stability, Instability and Chaos* (pp. 33–82). Springer International Publishing. https://doi.org/10.1007/978-3-319-56138-7_3 (accessed on 30th July 2025).

165. Othman, M., Kasap, E., & Korozlu, N. (2010). Ab-initio investigation of structural, electronic and optical properties of In⊠Ga$_1$–⊠As, GaAs$_1$–⊠P⊠ ternary and In⊠Ga$_1$–⊠As$_1$–⊠P⊠ quaternary semiconductor alloys. *Journal of Alloys and Compounds, 496*(1), 226–233. https://doi.org/10.1016/j.jallcom.2009.12.109 (accessed on 30th July 2025).

166. Pan, D., Xu, H., & García de Abajo, F. J. (2021). Rotational Doppler cooling and heating. *Science Advances, 7*(2), eabd6705. https://doi.org/10.1126/sciadv.abd6705 (accessed on 30th July 2025).

167. Panish, M. B., & Hayashi, I. (1974). Heterostructure junction lasers. In R. Wolfe (Ed.), *Applied Solid State Science* (Vol. 4, pp. 235–328). Elsevier. https://doi.org/10.1016/B978-0-12-002904-4.50011-5 (accessed on 30th July 2025).

168. Pawley, J. B. (2006). Points, pixels, and gray levels: Digitizing image data. In J. B. Pawley (Ed.), *Handbook of Biological Confocal Microscopy* (pp. 59–79). Springer US. https://doi.org/10.1007/978-0-387-45524-2_4 (accessed on 30th July 2025).

169. Pearton, S. J., Zolper, J. C., Shul, R. J., & Ren, F. (1999). GaN: Processing, defects, and devices. *Journal of Applied Physics, 86*(1), 1–78. https://doi.org/10.1063/1.371145 (accessed on 30th July 2025).

170. Pereira, R. N., & Almeida, A. J. (2015). Doped semiconductor nanoparticles synthesized in gas-phase plasmas. *Journal of Physics D: Applied Physics,*

48(31), 314005. https://doi.org/10.1088/0022-3727/48/31/314005 (accessed on 30th July 2025).

171. Pessa, M., Peng, C. S., Jouhti, T., Pavelescu, E. M., Li, W., Karirinne, S., Liu, H., & Okhotnikov, O. (2003). Long-wavelength nitride lasers on GaAs. Microelectronic Engineering, 69(2), 195–207. https://doi.org/10.1016/S0167-9317(03)00297-1 (accessed on 30th July 2025).

172. Peterka, P., & Vojtěch, J. (2024). Optical amplification. In Handbook of Radio and Optical Networks Convergence (pp. 91–141). Springer, Singapore. https://doi.org/10.1007/978-981-97-2282-2_20 (accessed on 30th July 2025).

173. Petrie, G., & Toth, C. K. (2018). Introduction to laser ranging, profiling, and scanning. In Topographic Laser Ranging and Scanning (2nd ed.). CRC Press. (accessed on 30th July 2025).

174. Pirouz, P. (2013). The concept of quasi-Fermi level and expansion of faulted loops in SiC under minority carrier injection. Physica Status Solidi (a), 210(1), 181–186. https://doi.org/10.1002/pssa.201200501 (accessed on 30th July 2025).

175. Pitz, G. A., & Anderson, M. D. (2017). Recent advances in optically pumped alkali lasers. Applied Physics Reviews, 4(4), 041101. https://doi.org/10.1063/1.5006913 (accessed on 30th July 2025).

176. Placidi, E., Arciprete, F., Magri, R., Rosini, M., Vinattieri, A., Cavigli, L., Gurioli, M., Giovine, E., Persichetti, L., Fanfoni, M., Patella, F., & Balzarotti, A. (2012). InAs epitaxy on GaAs(001): A model case of strain-driven self-assembling of quantum dots. In S. Bellucci (Ed.), Self-Assembly of Nanostructures: The INFN Lectures, Vol. III (pp. 73–125). Springer. https://doi.org/10.1007/978-1-4614-0742-3_2 (accessed on 30th July 2025).

177. Plekhanov, V. G. (2012). Electron excitations in low-dimensional structures. In V. G. Plekhanov (Ed.), Isotope Low-Dimensional Structures: Elementary Excitations and Applications (pp. 31–70). Springer. https://doi.org/10.1007/978-3-642-28613-1_3 (accessed on 30th July 2025).

178. Poon, A. W., Luo, X., Xu, F., & Chen, H. (2009). Cascaded microresonator-based matrix switch for silicon on-chip optical interconnection. Proceedings of the IEEE, 97(7), 1216–1238. https://doi.org/10.1109/JPROC.2009.2014884 (accessed on 30th July 2025).

179. Popovici, G., & Morkog, H. (2000). Growth and doping of defects in di-nitrides. In GaN and Related Materials II. CRC Press. (accessed on 30th July 2025).

180. Rahimi-Iman, A. (2016). Recent advances in VECSELs. Journal of Optics, 18(9), 093003. https://doi.org/10.1088/2040-8978/18/9/093003 (accessed on 30th July 2025).

181. Ramesh, Ch., Tyagi, P., Yadav, B. S., Ojha, S., Maurya, K. K., Senthil Kumar, M., & Kushvaha, S. S. (2018). Effect of nitridation temperature on formation and properties of GaN nanowall networks on sapphire (0001) grown by laser MBE. Materials Science and Engineering: B, 231, 105–114. https://doi.org/10.1016/j.mseb.2018.10.009 (accessed on 30th July 2025).

182. Rantamäki, A., & Guina, M. (2021). VECSELs in the wavelength range 1.18–1.55 µm. In Vertical External Cavity Surface Emitting Lasers (pp. 27–62). John Wiley & Sons, Ltd. https://doi.org/10.1002/9783527807956.ch2 (accessed on 30th July 2025).

183. Rao, Y., Yang, W., Chase, C., Huang, M. C. Y., Worland, D. P., Khaleghi, S., Chitgarha, M. R., Ziyadi, M., Willner, A. E., & Chang-Hasnain, C. J. (2013). Long-wavelength VCSEL using high-contrast grating. IEEE Journal of Selected Topics in Quantum Electronics, 19(4), 1701311–1701311. https://doi.org/10.1109/JSTQE.2013.2246780 (accessed on 30th July 2025).

184. Rediker, R. H. (1965). Semiconductor lasers. Physics Today, 18(2), 42–50. https://doi.org/10.1063/1.3047183 (accessed on 30th July 2025).

185. Reshchikov, M. A. (2023). On the origin of the yellow luminescence band in GaN. Physica Status Solidi (b), 260(8), 2200488. https://doi.org/10.1002/pssb.202200488 (accessed on 30th July 2025).

186. Ribeiro da Luz, B., & Crowley, J. K. (2007). Spectral reflectance and emissivity features of broad leaf plants: Prospects for remote sensing in the thermal infrared (8.0–14.0 µm). Remote Sensing of Environment, 109(4), 393–405. https://doi.org/10.1016/j.rse.2007.01.008 (accessed on 30th July 2025).

187. Ringel, S. A., & Grassman, T. J. (2011). 14III–V solar cells on silicon. In III-V Compound Semiconductors. CRC Press. (accessed on 30th July 2025).

188. Rønnekleiv, E. (2001). Frequency and intensity noise of single frequency fiber Bragg grating lasers. Optical Fiber Technology, 7(3), 206–235. https://doi.org/10.1006/ofte.2001.0357 (accessed on 30th July 2025).

189. Rousseau, B., Brun, J. F., Meneses, D. D. S., & Echegut, P. (2005). Temperature measurement: Christiansen wavelength and blackbody reference. International Journal of Thermophysics, 26(4), 1277–1286. https://doi.org/10.1007/s10765-005-6726-4 (accessed on 30th July 2025).

190. Salières, P., L'Huillier, A., Antoine, P., & Lewenstein, M. (1999). Study of the spatial and temporal coherence of high-order harmonics. In B. Bederson & H. Walther (Eds.), Advances in Atomic, Molecular, and Optical Physics (Vol. 41, pp. 83–142). Academic Press. https://doi.org/10.1016/S1049-250X(08)60219-0 (accessed on 30th July 2025).

191. Sameshima, T., Yasuta, K., Hasumi, M., Nagao, T., & Inouchi, Y. (2018). Activation of dopant in silicon by ion implantation under heating sample at 200 °C. Applied Physics A, 124(3), 228. https://doi.org/10.1007/s00339-018-1656-8 (accessed on 30th July 2025).

192. Samuel, I. D. W., & Turnbull, G. A. (2007). Organic semiconductor lasers. Chemical Reviews, 107(4), 1272–1295. https://doi.org/10.1021/cr050152i (accessed on 30th July 2025).

193. Sanguinetti, S., Bietti, S., & Koguchi, N. (2018). Chapter 13—Droplet epitaxy of nanostructures. In M. Henini (Ed.), Molecular Beam Epitaxy (Second edition, pp. 293–314). Elsevier. https://doi.org/10.1016/B978-0-12-812136-8.00013-X (accessed on 30th July 2025).

194. Sarma, K. (2015). Recent progress in OLED and flexible displays and their potential for application to aerospace and military display systems. Display Technologies and Applications for Defense, Security, and Avionics IX; and Head- and Helmet-Mounted Displays XX, 9470, 128–138. https://doi.org/10.1117/12.2180172 (accessed on 30th July 2025).

195. Sato, H., Toth, R. S., & Honjo, G. (1967). Long period stacking order in close packed structures of metals. Journal of Physics and Chemistry of Solids, 28(2), 137–160. https://doi.org/10.1016/0022-3697(67)90104-7 (accessed on 30th July 2025).

196. Shallcross, R. C., Olthof, S., Meerholz, K., & Armstrong, N. R. (2019). Impact of titanium dioxide surface defects on the interfacial composition and energetics of evaporated perovskite active layers. ACS Applied Materials & Interfaces, 11(35), 32500–32508. https://doi.org/10.1021/acsami.9b09935 (accessed on 30th July 2025).

197. Shayeganrad, G. (2020). An introduction to laser. In S. Stübinger, F. Klämpfl, M. Schmidt, & H. F. Zeilhofer (Eds.), Lasers in Oral and Maxillofacial Surgery (pp. 9–23). Springer International Publishing. https://doi.org/10.1007/978-3-030-29604-9_2 (accessed on 30th July 2025).

198. Shchekin, O., & Craford, M. G. (2016). History of solid-state light sources. In Handbook of Advanced Lighting Technology (pp. 1–30). Springer, Cham. https://doi.org/10.1007/978-3-319-00295-8_63-1 (accessed on 30th July 2025).

199. Shen, B. (2015). Development trends of GaN-based wide bandgap semiconductors: From solid state lighting to power electronic devices. Frontiers of Optoelectronics, 8(4), 456–460. https://doi.org/10.1007/s12200-015-0459-1 (accessed on 30th July 2025).

200. Sinusia Lozano, M., & Gómez, V. J. (2023). Epitaxial growth of crystal phase quantum dots in III–V semiconductor nanowires. Nanoscale Advances, 5(7), 1890–1909. https://doi.org/10.1039/D2NA00956K (accessed on 30th July 2025).

201. Smith, A. M., & Nie, S. (2010). Semiconductor nanocrystals: Structure, properties, and band gap engineering. Accounts of Chemical Research, 43(2), 190–200. https://doi.org/10.1021/ar9001069 (accessed on 30th July 2025).

202. Sneh, A., & Doerr, C. R. (1999). Indium phosphide-based photonic circuits and components. In Integrated Optical Circuits and Components. CRC Press. (accessed on 30th July 2025).

203. Socorro, A. B., Hernaez, M., Del Villar, I., Corres, J. M., Arregui, F. J., & Matias, I. R. (2016). Single-mode—multimode—single-mode and lossy mode resonance-based devices: A comparative study for sensing applications. Microsystem Technologies, 22(7), 1633–1638. https://doi.org/10.1007/s00542-015-2793-z (accessed on 30th July 2025).

204. Soh, C. B., Chua, S. J., Lim, H. F., Chi, D. Z., Tripathy, S., & Liu, W. (2004). Assignment of deep levels causing yellow luminescence in GaN. Journal of Applied Physics, 96(3), 1341–1347. https://doi.org/10.1063/1.1757654 (accessed on 30th July 2025).

205. Song, Q., Fang, W., Liu, B., Ho, S. T., Solomon, G. S., & Cao, H. (2009). Chaotic microcavity laser with high quality factor and unidirectional output. Physical Review A, 80(4), 041807. https://doi.org/10.1103/PhysRevA.80.041807 (accessed on 30th July 2025).

206. Spear, W. E. (1977). Doped amorphous semiconductors. Advances in Physics, 26(6), 811–845. https://doi.org/10.1080/00018737700101463 (accessed on 30th July 2025).

207. Spink, C. H. (2008). Differential scanning calorimetry. In Methods in Cell Biology (Vol. 84, pp. 115–141). Academic Press. https://doi.org/10.1016/S0091-679X(07)84005-2 (accessed on 30th July 2025).

208. Srivastava, A. K. (1990). Semiconductor lasers for optical communication. Bulletin of Materials Science, 13(1), 151–159. https://doi.org/10.1007/BF02744868 (accessed on 30th July 2025).

209. Stoltz, M. P., Burgaud, P., Murgadella, F., Hirtz, J. P., Petit, P., & Vervoitte, A. (1998). Reliability and qualification methodology of 60 W QCW linear

bar arrays. Microelectronics Reliability, 38(4), 689–696. https://doi.org/10.1016/S0026-2714(97)00202-3 (accessed on 30th July 2025).

210. Stone, B. T., & Bray, K. L. (1996). Fluorescence properties of Er^{3+}-doped sol–gel glasses. Journal of Non-Crystalline Solids, 197(2), 136–144. https://doi.org/10.1016/0022-3093(95)00514-5 (accessed on 30th July 2025).

211. Strickland, D. (2021). Chirped pulse amplification. In Handbook of Laser Technology and Applications (2nd ed.). CRC Press. (accessed on 30th July 2025).

212. Sturm, K. (1982). Electron energy loss in simple metals and semiconductors. Advances in Physics, 31(1), 1–64. https://doi.org/10.1080/00018738200101348 (accessed on 30th July 2025).

213. Suematsu, Y. (1983). Long-wavelength optical fiber communication. Proceedings of the IEEE, 71(6), 692–721. https://doi.org/10.1109/PROC.1983.12659 (accessed on 30th July 2025).

214. Sun, W. H., Wang, L. S., Chua, S. J., Chen, K. M., & Qin, G. G. (2001). Local vibrational modes in gamma-irradiated GaN grown by metal-organic chemical vapor deposition. Materials Science in Semiconductor Processing, 4(6), 559–562. https://doi.org/10.1016/S1369-8001(02)00017-3 (accessed on 30th July 2025).

215. Svelto, O. (2010). Interaction of radiation with atoms and ions. In O. Svelto (Ed.), Principles of Lasers (pp. 17–79). Springer US. https://doi.org/10.1007/978-1-4419-1302-9_2 (accessed on 30th July 2025).

216. Tansley, T. L., Goldys, E. M., Goldlewski, M., Zhou, B., & Zuo, H. Y. (1997). The contribution of defects to the electrical and optical properties of GaN. In GaN and Related Materials. CRC Press. (accessed on 30th July 2025).

217. Tassev, V. L., & Vangala, S. R. (2019). Thick hydride vapor phase heteroepitaxy: A novel approach to growth of nonlinear optical materials. Crystals, 9(8), Article 8. https://doi.org/10.3390/cryst9080393 (accessed on 30th July 2025).

218. Thaler, G. T., Koleske, D. D., Lee, S. R., Bogart, K. H. A., & Crawford, M. H. (2010). Thermal stability of thin InGaN films on GaN. Journal of Crystal Growth, 312(11), 1817–1822. https://doi.org/10.1016/j.jcrysgro.2010.03.008 (accessed on 30th July 2025).

219. Thyagarajan, K., & Ghatak, A. (2011). Semiconductor lasers. In K. Thyagarajan & A. Ghatak (Eds.), Lasers: Fundamentals and Applications (pp. 323–362). Springer US. https://doi.org/10.1007/978-1-4419-6442-7_13 (accessed on 30th July 2025).

220. Tingberg, T., Ive, T., & Larsson, A. (2017). Investigation of Si and O donor impurities in unintentionally doped MBE-grown GaN on SiC(0001) substrate. Journal of Electronic Materials, 46(8), 4898–4902. https://doi.org/10.1007/s11664-017-5484-y (accessed on 30th July 2025).

221. Tobing, L. Y. M., & Dumon, P. (2010). Fundamental principles of operation and notes on fabrication of photonic microresonators. In I. Chremmos, O. Schwelb, & N. Uzunoglu (Eds.), Photonic Microresonator Research and Applications (pp. 1–27). Springer US. https://doi.org/10.1007/978-1-4419-1744-7_1 (accessed on 30th July 2025).

222. Tong, X. C. (2014). Optoelectronic devices integrated with optical waveguides. In X. C. Tong Ph.D (Ed.), Advanced Materials for Integrated Optical Waveguides (pp. 103–160). Springer International Publishing. https://doi.org/10.1007/978-3-319-01550-7_3 (accessed on 30th July 2025).

223. Tsang, W. T. (1987). Chapter 7 quantum confinement heterostructure semiconductor lasers. In R. Dingle (Ed.), Semiconductors and Semimetals (Vol. 24, pp. 397–458). Elsevier. https://doi.org/10.1016/S0080-8784(08)62454-0 (accessed on 30th July 2025).

224. Tull, J. X., Dugan, M. A., & Warren, W. S. (1997). High-resolution, ultrafast laser pulse shaping and its applications. In W. S. Warren (Ed.), Advances in Magnetic and Optical Resonance (Vol. 20, pp. 1–II). Academic Press. https://doi.org/10.1016/S1057-2732(97)80002-2 (accessed on 30th July 2025).

225. Vardi, A., Abrashkevich, D., Frishman, E., & Shapiro, M. (1997). Theory of radiative recombination with strong laser pulses and the formation of ultracold molecules via stimulated photo-recombination of cold atoms. The Journal of Chemical Physics, 107(16), 6166–6174. https://doi.org/10.1063/1.474282 (accessed on 30th July 2025).

226. Vragović, I., & Scholz, R. (2003). Frenkel exciton model of optical absorption and photoluminescence in α-PTCDA. Physical Review B, 68(15), 155202. https://doi.org/10.1103/PhysRevB.68.155202 (accessed on 30th July 2025).

227. Wang, H., Zhang, L., Chen, Z., Hu, J., Li, S., Wang, Z., Liu, J., & Wang, X. (2014). Semiconductor heterojunction photocatalysts: Design, construction, and photocatalytic performances. Chemical Society Reviews, 43(15), 5234–5244. https://doi.org/10.1039/C4CS00126E (accessed on 30th July 2025).

228. Wang, J., Li, X., Hou, D., Feng, F., Liu, Y., & Liu, X. (2015). Advances in bonding technology for high power diode laser bars.

Components and Packaging for Laser Systems, 9346, 30–50. https://doi.org/10.1117/12.2079619 (accessed on 30th July 2025).

229. Wang, J., Wang, X., Chen, J., Gao, X., Zeng, X., Mao, H., & Xu, K. (2021). Investigation on minority carrier lifetime, diffusion length and recombination mechanism of Mg-doped GaN grown by MOCVD. Journal of Alloys and Compounds, 870, 159477. https://doi.org/10.1016/j.jallcom.2021.159477 (accessed on 30th July 2025).

230. Wang, J., Yuan, Z., Guo, L., Xiong, L., Zhang, Y., Peng, C., Li, X., & Liu, X. (2010). Packaging of high-power semiconductor laser arrays using a novel macro-channel cooler. 2010 11th International Conference on Electronic Packaging Technology & High-Density Packaging, 92–97. https://doi.org/10.1109/ICEPT.2010.5582358 (accessed on 30th July 2025).

231. Wang, X., & Yoshikawa, A. (2004). Molecular beam epitaxy growth of GaN, AlN and InN. Progress in Crystal Growth and Characterization of Materials, 48–49, 42–103. https://doi.org/10.1016/j.pcrysgrow.2005.03.002 (accessed on 30th July 2025).

232. Wang, Z., Li, T., Yang, G., & Song, Y. (2017). High power, high efficiency continuous-wave 808 nm laser diode arrays. Optics & Laser Technology, 97, 297–301. https://doi.org/10.1016/j.optlastec.2017.07.015 (accessed on 30th July 2025).

233. Wei, S., Gao, X., Wang, X., Pan, Y., Zeng, X., Chen, J., Zheng, S., & Xu, K. (2023). Influence of polarities on optical properties of Mg-doped GaN films grown on GaN free-standing substrates by MOCVD. Journal of Luminescence, 257, 119740. https://doi.org/10.1016/j.jlumin.2023.119740 (accessed on 30th July 2025).

234. Wei, X., Zhang, P., Xu, T., Zhou, H., Bai, Y., & Chen, Q. (2022). Chemical approaches for electronic doping in photovoltaic materials beyond crystalline silicon. Chemical Society Reviews, 51(24), 10016–10063. https://doi.org/10.1039/D2CS00110A (accessed on 30th July 2025).

235. Wenzel, H., & Zeghuzi, A. (2017). High-power lasers. In Handbook of Optoelectronic Device Modeling and Simulation. CRC Press. (accessed on 30th July 2025).

236. Wenzel, H., Kantner, M., Radziunas, M., & Bandelow, U. (2021). Semiconductor laser linewidth theory revisited. Applied Sciences, 11(13), Article 13. https://doi.org/10.3390/app11136004 (accessed on 30th July 2025).

237. Wiersig, J., Kim, S. W., & Hentschel, M. (2008). Asymmetric scattering and nonorthogonal mode patterns in optical microspirals. Physical Review

A, 78(5), 053809. https://doi.org/10.1103/PhysRevA.78.053809 (accessed on 30th July 2025).

238. Wysocki, G., Lewicki, R., Curl, R. F., Tittel, F. K., Diehl, L., Capasso, F., Troccoli, M., Hofler, G., Bour, D., Corzine, S., Maulini, R., Giovannini, M., & Faist, J. (2008). Widely tunable mode-hop free external cavity quantum cascade lasers for high resolution spectroscopy and chemical sensing. Applied Physics B, 92(3), 305–311. https://doi.org/10.1007/s00340-008-3047-x (accessed on 30th July 2025).

239. Xu, B., Qu, K., Wang, Z., Ma, S., Liang, J., Jia, Z., Dong, H., Liu, X., Wong, W. Y., & Wu, Y. (2020). Investigation of photoelectric performance of laser diode by regulation of p-waveguide layer thickness. Optik, 200, 163458. https://doi.org/10.1016/j.ijleo.2019.163458 (accessed on 30th July 2025).

240. Xu, C., Dai, J., Zhu, G., Zhu, G., Lin, Y., Li, J., & Shi, Z. (2014). Whispering-gallery mode lasing in ZnO microcavities. Laser & Photonics Reviews, 8(4), 469–494. https://doi.org/10.1002/lpor.201300127 (accessed on 30th July 2025).

241. Xu, K., Wan, P., Liu, M., Tang, K., Li, L., He, T., Shi, D. N., Kan, C., & Jiang, M. (2024). High Q-factor and low threshold electrically pumped single-mode microlaser based on a single-microwire double-heterojunction device. ACS Photonics, 11(8), 3276–3287. https://doi.org/10.1021/acsphotonics.4c00667 (accessed on 30th July 2025).

242. Yabana, K., Otobe, T., & Iwata, J. I. (2006). First-principles density-functional approach for many-electron dynamics under intense laser fields. In K. Yamanouchi, S. L. Chin, P. Agostini, & G. Ferrante (Eds.), Progress in Ultrafast Intense Laser Science Volume I (pp. 77–94). Springer. https://doi.org/10.1007/3-540-34422-5_5 (accessed on 30th July 2025).

243. Yabre, G. (1996). Effect of relatively strong light injection on the chirp-to-power ratio and the 3 dB bandwidth of directly modulated semiconductor lasers. Journal of Lightwave Technology, 14(10), 2367–2373. https://doi.org/10.1109/50.541231 (accessed on 30th July 2025).

244. Yakuhina, A. V., Kadochkin, A. S., Gorelov, D. V., Svetukhin, V. V., Generalov, S. S., & Amelichev, V. V. (2021). Influence of the surface roughness of a silicon disk resonator on its Q-factor. Photonics, 8(6), Article 6. https://doi.org/10.3390/photonics8060225 (accessed on 30th July 2025).

245. Yamada, M. (2014). Overview of the operating principles of lasers. In M. Yamada (Ed.), Theory of Semiconductor Lasers: From Basis of Quantum Electronics to Analyses of the Mode Competition Phenomena and Noise

(pp. 3–15). Springer Japan. https://doi.org/10.1007/978-4-431-54889-8_1 (accessed on 30th July 2025).

246. Yamamoto, Y., Machida, S., & Björk, G. (1992). Micro-cavity semiconductor lasers with controlled spontaneous emission. Optical and Quantum Electronics, 24(2), S215–S243. https://doi.org/10.1007/BF00625826 (accessed on 30th July 2025).

247. Yan, J., Yuan, J., Jiang, Y., Zhu, H., Choi, H. W., & Wang, Y. (2022). A vertical AlGaN DUV light-emitting diode fabricated by wafer bonding and sapphire thinning technology. Applied Physics Express, 15(3), 032003. https://doi.org/10.35848/1882-0786/ac4e24 (accessed on 30th July 2025).

248. Yan, Z. H., & Zhou, S. (2019). Bonding stress and reliability of low-polarization quantum-well superluminescent diode. Physica E: Low-Dimensional Systems and Nanostructures, 109, 140–143. https://doi.org/10.1016/j.physe.2019.01.013 (accessed on 30th July 2025).

249. Yang, M., & Lyu, D. (2020). Optical fiber sensor set-up elements. In Optical Fiber Sensors (pp. 49–89). John Wiley & Sons, Ltd. https://doi.org/10.1002/9781119534730.ch3 (accessed on 30th July 2025).

250. Yasui, K. (2018). Acoustic cavitation. In K. Yasui (Ed.), Acoustic Cavitation and Bubble Dynamics (pp. 1–35). Springer International Publishing. https://doi.org/10.1007/978-3-319-68237-2_1 (accessed on 30th July 2025).

251. Ye, J., Guan, C., Lv, P., Wang, W., Chen, X., Wang, Z., Xiao, Y., Zhan, L., Gong, J., & Yao, Y. (2024). 780 nm narrow linewidth external cavity diode laser for quantum sensing. Sensors, 24(22), Article 22. https://doi.org/10.3390/s24227237 (accessed on 30th July 2025).

252. Yee, K. S., & Chen, J. S. (1997). The finite-difference time-domain (FDTD) and the finite-volume time-domain (FVTD) methods in solving Maxwell's equations. IEEE Transactions on Antennas and Propagation, 45(3), 354–363. https://doi.org/10.1109/8.558651 (accessed on 30th July 2025).

253. Yeh, C. H., Hsu, Y., & Chow, C. W. (2016). Utilizing a silicon-photonic micro-ring-resonator and multi-ring scheme for wavelength-switchable erbium fiber laser in single-longitudinal-mode. Laser Physics Letters, 13(6), 065103. https://doi.org/10.1088/1612-2011/13/6/065103 (accessed on 30th July 2025).

254. Yokoyama, H., Nishi, K., Anan, T., Nambu, Y., Brorson, S. D., Ippen, E. P., & Suzuki, M. (1992). Controlling spontaneous emission and threshold-less laser oscillation with optical microcavities. Optical and Quantum Electronics, 24(2), S245–S272. https://doi.org/10.1007/BF00625827 (accessed on 30th July 2025).

255. Yoshie, T., Tang, L., & Su, S. Y. (2011). Optical microcavity: Sensing down to single molecules and atoms. Sensors, 11(2), Article 2. https://doi.org/10.3390/s110201972 (accessed on 30th July 2025).

256. Yoshikawa, A., Matsunami, H., & Nanishi, Y. (2007). Development and applications of wide bandgap semiconductors. In K. Takahashi, A. Yoshikawa, & A. Sandhu (Eds.), Wide Bandgap Semiconductors: Fundamental Properties and Modern Photonic and Electronic Devices (pp. 1–24). Springer. https://doi.org/10.1007/978-3-540-47235-3_1 (accessed on 30th July 2025).

257. Zappa, F., Tisa, S., Tosi, A., & Cova, S. (2007). Principles and features of single-photon avalanche diode arrays. Sensors and Actuators A: Physical, 140(1), 103–112. https://doi.org/10.1016/j.sna.2007.06.021 (accessed on 30th July 2025).

258. Zappe, H. (2004a). Laser diode structures. In H. Zappe (Ed.), Laser Diode Microsystems (pp. 115–159). Springer. https://doi.org/10.1007/978-3-662-08249-2_4 (accessed on 30th July 2025).

259. Zappe, H. (2004b). Semiconductors for lasers. In H. Zappe (Ed.), Laser Diode Microsystems (pp. 63–114). Springer. https://doi.org/10.1007/978-3-662-08249-2_3 (accessed on 30th July 2025).

260. Zatsepin, D. A., Zatsepin, A. F., Boukhvalov, D. W., Kurmaev, E. Z., Pchelkina, Z. V., & Gavrilov, N. V. (2016). Electronic structure and photoluminescence properties of Zn-ion implanted silica glass before and after thermal annealing. Journal of Non-Crystalline Solids, 432, 183–188. https://doi.org/10.1016/j.jnoncrysol.2015.10.002 (accessed on 30th July 2025).

261. Zhang, J., Li, K., Zhang, T., Buenconsejo, P. J. S., Chen, M., Wang, Z., Zhang, M., Wang, Z., & Wei, L. (2017). Laser-induced in-fiber fluid dynamical instabilities for precise and scalable fabrication of spherical particles. Advanced Functional Materials, 27(43), 1703245. https://doi.org/10.1002/adfm.201703245 (accessed on 30th July 2025).

262. Zhang, Z., Zhang, P., Li, X., Xiong, L., Liu, H., Nie, Z., Wang, Z., & Liu, X. (2012). Thermal modeling and analysis of high-power semiconductor laser arrays. 2012 13th International Conference on Electronic Packaging Technology & High-Density Packaging, 560–566. https://doi.org/10.1109/ICEPT-HDP.2012.6474681 (accessed on 30th July 2025).

263. Zhao, Q., Pei, L., Wang, J., Xie, Y., Ruan, Z., Zheng, J., Li, J., & Ning, T. (2021). Interval-adjustable multi-wavelength erbium-doped fiber laser with the assistance of NOLM or NALM. IEEE Access, 9, 16316–16322. https://doi.org/10.1109/ACCESS.2021.3050462 (accessed on 30th July 2025).

264. Zhao, X., Montgomery, K. H., & Woodall, J. M. (2014). Hall effect studies of AlGaAs grown by liquid-phase epitaxy for tandem solar cell applications. Journal of Electronic Materials, 43(11), 3999–4002. https://doi.org/10.1007/s11664-014-3340-x (accessed on 30th July 2025).

265. Zhong, Y., Liao, K., Du, W., Zhu, J., Shang, Q., Zhou, F., Wu, X., Sui, X., Shi, J., Yue, S., Wang, Q., Zhang, Y., Zhang, Q., Hu, X., & Liu, X. (2020). Large-scale thin $CsPbBr_3$ single-crystal film grown on sapphire via chemical vapor deposition: Toward laser array application. ACS Nano, 14(11), 15605–15615. https://doi.org/10.1021/acsnano.0c06380 (accessed on 30th July 2025).

266. Zhou, H., Lee, J., Kang, M., Kim, H., Lee, H., & In, J. B. (2022). All laser-based fabrication of microchannel heat sink. Materials & Design, 221, 110968. https://doi.org/10.1016/j.matdes.2022.110968 (accessed on 30th July 2025).

267. Zhou, K. J., & Zhang, X. L. (2010). Multiwavelength erbium-doped fiber ring laser using a $LiNbO_3$ multifunction chip for fiber gyroscope. Laser Physics, 20(6), 1428–1432. https://doi.org/10.1134/S1054660X1011037X (accessed on 30th July 2025).

268. Zhou, W., Bandyopadhyay, N., Wu, D., McClintock, R., & Razeghi, M. (2016). Monolithically, widely tunable quantum cascade lasers based on a heterogeneous active region design. Scientific Reports, 6(1), 25213. https://doi.org/10.1038/srep25213 (accessed on 30th July 2025).

269. Zhu, Q., Karlsson, K. F., Pelucchi, E., & Kapon, E. (2007). Transition from two-dimensional to three-dimensional quantum confinement in semiconductor quantum wires/quantum dots. Nano Letters, 7(8), 2227–2233. https://doi.org/10.1021/nl0706650 (accessed on 30th July 2025).

270. Zogg, H., & Ishida, A. (2001). IV-VI (lead chalcogenide) infrared sensors and lasers. In P. Capper & C. T. Elliott (Eds.), Infrared Detectors and Emitters: Materials and Devices (pp. 43–75). Springer US. https://doi.org/10.1007/978-1-4615-1607-1_3 (accessed on 30th July 2025).

271. Zuo, G., Abdalla, H., & Kemerink, M. (2016). Impact of doping on the density of states and the mobility in organic semiconductors. Physical Review B, 93(23), 235203. https://doi.org/10.1103/PhysRevB.93.235203 (accessed on 30th July 2025).

INDEX

Symbols

A

B

C

D

E

F

G

H

T

U

V